Society and Environment

Society and Environment

SC Naik PhD

Former Dean and Professor and Head
Chemical Engineering
National Institute of Technology
Rourkela, Orissa, India

TN Tiwari PhD

Fromer Professor and Head
Department of Physics
National Institute of Technology
Rourkela, Orissa, India

Oxford & IBH Publishing Co. Pvt. Ltd.
New Delhi
(A Unit of CBS Publishers & Distributors Pvt Ltd)

CBS

CBS Publishers & Distributors Pvt Ltd

New Delhi • Bengaluru • Chennai • Kochi • Kolkata • Mumbai
Hyderabad • Jharkhand • Nagpur • Patna • Pune • Uttarakhand

Society and Environment

ISBN-13: 978-81-204-1659-8
ISBN-10: 81-204-1659-7 .

OXFORD & IBH

New Delhi
(A Unit of CBS Publishers & Distributors Pvt Ltd)

Published by Satish Kumar Jain and Produced by Varun Jain for

CBS Publishers & Distributors Pvt Ltd
4819/XI Prahlad Street, 24 Ansari Road, Daryaganj, New Delhi 110 002, India.

Ph: 23289259, 23266861, 23266867   Fax: 011-23243014   Website: www.cbspd.com
                                                             e-mail: delhi@cbspd.com;
                                                             cbspubs@airtelmail.in.

Corporate Office: 204 FIE, Industrial Area, Patparganj, Delhi 110 092, India
Ph: 4934 4934   Fax: 4934 4935   e-mail: publishing@cbspd.com;
                                                             publicity@cbspd.com

Branches

- **Bengaluru:** Seema House 2975, 17th Cross, K.R. Road, Banasankari 2nd Stage, Bengaluru 560 070, Karnataka
  Ph: +91-80-26771678/79   Fax: +91-80-26771680   e-mail: bangalore@cbspd.com
- **Chennai:** 7, Subbaraya Street, Shenoy Nagar, Chennai 600 030, Tamil Nadu
  Ph: +91-44-26680620, 26681266   Fax: +91-44-42032115   e-mail: chennai@cbspd.com
- **Kochi:** Ashana House, 39/1904, AM Thomas Road, Valanjambalam, Ernakulam 682 016, Kochi, Kerala
  Ph: +91-484-4059061-65,67   Fax: +91-484-4059065   e-mail: kochi@cbspd.com
- **Kolkata:** 6/B, Ground Floor, Rameswar Shaw Road, Kolkata-700014 (West Bengal), India
  Ph: +91-33-2289-1126, 2289-1127, 2289-1128   e-mail: kolkata@cbspd.com
- **Mumbai:** 83-C, Dr E Moses Road, Worli, Mumbai-400018, Maharashtra
  Ph: +91-22-24902340/41   Fax: +91-22-24902342   e-mail: mumbai@cbspd.com

Representatives

• **Hyderabad**	0-9885175004	• **Jharkhand**	0-9811541605	• **Nagpur**	0-9021734563
• **Patna**	0-9334159340	• **Pune**	0-9623451994	• **Uttarakhand**	0-9716462459

Printed at Chaman Enterprises, Daryaganj, New Delhi, India

Dedicated to
our
Society and Environment

Preface

Environment simply means surroundings. Society and Environment are inseparable and are utterly interdependent. Knowledge of both are essential and a definite place has to be found for a course on Society and Environment in the academy curriculum. Their relative importance in a given course is a matter of judgment. However, one should bear in mind that a good treatment of Environment can only follow a sound understanding of Society. Study of one without the other may be a serious lapse. Realising such an importance an Expert Committee constituted by the Institution of Engineers (India), the largest professional body in the country, prepared a balanced Syllabus for its AMIE examinations which was subsequently approved by the MHRD and the AICTE. We were requested by the Institution to write a suitable textbook based on the approved Syllabus.

This book is not intended for advanced students, nor have we attempted to give an all-inclusive treatment of the topics we have selected. Because of the express attempt to write a textbook more or less within the compass of an academic session or a semester, it has been necessary to omit many aspects. The purpose of the book is to present to the students an account of the fundamentals of the subject. Although the text contains much introductory materials, it is also reasonably comprehensive. Every chapter contains typical questions for students of all levels. The entire book (Society and Environment) or one Group (Society or Environment) can be covered in one academic session and when compressed, in one semester.

Many persons have contributed directly or indirectly to this book. One of us (SCN) had several interactions with many experts in USA, Canada and Singapore during his recent visits to these countries. We have referred to numerous publications. All these are gratefully acknowledged. Whatever be the shortcomings of this book, we hope that it will be found useful as a textbook at all levels of education. Comments and suggestions are welcome for effecting improvement in the next edition. We are indebted to the Institution of Engineers (India) for providing support to prepare the manuscript. We are thankful to every member of in our respective family for having shared our pain and pleasure. Sri S. K. Dash, National Institute of Technology, Rourkela, Orissa deserves our thanks for providing Secretarial assistance.

Incidentally, the Supreme Court has given a visionary direction that from the academic year 2004-05, Environment as a compulsory subject should be taught at all levels of education in our country. The NCERT for the school level, the UGC for the college level and the AICTE for the professional level have been assigned the responsibility. Through appropriate selection of chapters and sections, the Environment Group of this book can be used at all levels.

January, 2005
Rourkela, Orissa, India

S.C. NAIK
T.N. TIWARI

Syllabus

SOCIETY

Societal Structures and Dynamics

An analysis of basic sociological concepts and their application to contemporary society; social stratification, caste, class, cultural heritage, occupation, mobility and income distribution. Social tensions and their causes; societal responsibilities and social institutions.

Development Processes

Parameters for development. Interrelationship between social, economic and scientific factors. Role of science and technology in development. Planning – its objectives and assessment.

Technology Assessment

Historical development of science and technology. Criteria for assessment of appropriate technology and technology adaptation.

ENVIRONMENT

Ecosystems

Natural ecosystems. Principles of ecobalance. Biosphere cycle, carbon dioxide cycle. Causes for eco-imbalance – its effects and remedies.

Environmental Degradation

Causes for degradation – its effects. Control of air, water, soil and noise pollutions. Protection of ozone layer.

Waste Management

Agricultural, urban and industrial wastes.

Sustainable Development

Definition and concept. Technology for sustainable energy and materials.

Contents

GROUP - I

Society

Sociology and Its Scope

1.1 INTRODUCTION

Sociology is the study of human society and its dynamics. The main focus of the science of Sociology is the *Group* and not the individual. Sociology is the youngest branch of the Social Sciences. In fact it was the French philosopher **Auguste Comte** (1798 – 1857) who in 1842, coined the term "**sociology**" from the Latin word *Societas* (meaning "society") and the Greek word *logos* (meaning "the study" or " the science"). **Comte** defined Sociology as "the science of social phenomena, subject to natural invariable laws, the discovery of which is the object of investigation". Based on his definition **Comte** advocated the use of "*positive*" method for Sociology as is done in Natural Sciences like Physics since he believed that the scientific study of Sociology was of vital importance for the improvement of society.

1.2 DEFINITIONS OF SOCIOLOGY

Basically, Sociology is the study of or science of human society. The word "Sociology" has been defined by a number of eminent Sociologists and Philosophers in many different ways in order to emphasise different aspects of Sociology. As a result, no single definition of the term "Sociology" has been accepted as complete satisfactory. **Max Weber** (1864 –1920) defined Sociology as a science which interprets the social behavior with the aim of arriving at a causal explanation of human behavior. Italian Social Scientist **Pareto** maintained that Sociology is a logico-experimental science which studies the real individuals and real social phenomena. **Frederic Le Play** (1806 – 1882) held that Sociology is an observational science which studies actual social phenomena. It is interesting to note that **Le Play** practiced what he preached and as a result, he was the originator of *modern social survey*, which is one of the most important tools of sociological research. **L. T. Hobhouse** maintained that Sociology is a comprehensive science of society which aims at the interpretation of social life. According to **Morris Ginsberg,** Sociology aims at the study of the conditions and the consequences of human interactions and interrelations. According to **Ogburn,** Sociology is the science which deals with the study of social life of man and its relation with the factors concerning his cultural and natural environments. **Harry M. Johnson** defined Sociology as the branch of social science which deals with the study of social groups. According to **F. H. Giddings,** Sociology is the science of the association of mind and **Sorokin** defined Sociology as a generalized theory of the structure and dynamics of social systems. **MacIver** defined Sociology as the study of the nature of social relationships. According to him, Sociology is the study of social relationships among individuals and it aims at the discovery of the principles of cohesion and harmony in a social structure which conditions the activities of individuals. **Talcott Parsons** held the view that Sociology studies the structure and process in the social systems in which there are direct or indirect interactions amongst individuals. One of the briefest definitions of Sociology was given by **E. S. Bogardus** who defined Sociology as the study of social processes.

It is obvious from the definitions of Sociology given by various renowned Sociologists that Sociology is essentially the study of social relationships. Thus, we may briefly define sociology as *"a social science, which makes a scientific study of man's relationships in the society"*. A society is a web of social relationships and individuals held together in a group of individuals through social relations. Also, a society refers to the complex pattern of interactions which arise out of the relationship among individuals.

1.3 SCOPE OF SOCIOLOGY

1.3.1 Two Schools of Thoughts

There are mainly two schools of thoughts regarding the subject matter and scope of Sociology: (1) specialistic and (2) synthetic.

According to the *specialistic* school, the subject matter of Sociology consists of the forms of social relationships. The Sociologists belonging to this school regard Sociology as a pure science and they want to keep the scope of Sociology distinct from other social sciences. According to **Georg Simmel,** Sociology should confine itself to the study of the *formal* behaviour instead of the actual behaviour. Thus, Sociology comprehends the *forms* of social relationships and activities, and not the relationships themselves. According to him, Sociology is a *specific* social science which describes, classifies, and analyses the forms of relationships, the processes of socialisations and social organisations. Hence, the scope of Sociology includes the forms of human relationships and social processes. In contrast to this, the *synthetic* school of thought wants to make Sociology a *synthesis* of all the social sciences. Some of the prominent Sociologists belonging to this school are **Durkheim, Hobhouse** and **Sorokin.** According to them, Sociology is the science of sciences and, therefore, all the social sciences are included in its scope. All the aspects of social life are interrelated and the study of one aspect of the society cannot suffice to help us understand the society. For this reason, Sociologist should make a systematic study of the social life as a whole.

1.3.2 Wider Scope of Sociology

The scope of Sociology differs from other social sciences because Sociology mainly studies social relationships, but the study of social relationships makes it necessary to study other social sciences too. For example, if we want to investigate the causes of inter-caste tension from the sociological point of view, we have to seek the help of other relevant social sciences like History, Economics and Psychology. It is evident, therefore, that the wider scope of Sociology includes the subject matter of other social sciences. In other words although Sociology is distinct from other social sciences, it also synthesizes them for the purpose of an integrated study of the society.

According to **Morris Ginsberg,** Sociology studies four main problems:

(1) *Social Morphology* (study of human population and various types of social groups and institutions).
(2) *Social Control* (study of nature of laws, morals, conventions and other regulating agencies of the society).
(3) *Social Process* (study of various modes of interactions among individuals and groups of people).
(4) *Social Pathology* (study of various forms of maladjustments among individuals in social relations).

We may conclude the above discussion on the scope of Sociology by saying that Sociology, in its wider sense, studies all branches and all aspects of social phenomena.

1.4 SPECIALISED FIELDS OF SOCIOLOGY

There are four main objectives of sociological studies:

(1) To study the nature and functions of various social groups and the laws governing their development.

(2) To find out, by means of the comparative method, the interrelationship among various institutions.
(3) To formulate the laws (or empirical generalisations) of social development.
(4) To interpret the social laws in the light of more general laws of life.

Since the scope of Sociology is so vast and wide, Sociology has developed a large number of specialised fields.

Some of the major fields of Sociology are :
(1) Urban Sociology.
(2) Rural Sociology.
(3) Human Ecology.
(4) Demography.
(5) Industrial Sociology.
(6) Political Sociology.
(7) Criminology.
(8) Educational Sociology.
(9) Social Psychology.
(10) Sociology of Development.
(11) Social Disorganisation.
(12) Sociology of Religion.
(13) Medical Sociology.

Due to rapid advances in sociological research, many of the above-specialized fields are rapidly gaining the status of independent disciplines of study like Demography and Criminology.

1.5 SOCIOLOGY AND OTHER SOCIAL SCIENCES

The main objective of sociological research is to develop knowledge and skills that would help us to understand society, to assess the social implications of various changes that take place in the society as well as to plan and guide the social response to these changes. By its very nature, Sociology is apt to touch the domains of many other social sciences, which are clearly evident from the list of major fields of Sociology mentioned earlier. We discuss below briefly the relationship between Sociology and a few major social sciences:

1.5.1 Sociology and Economics

Economics has often been defined as the study of mankind in the ordinary business of life. Economics is considered as the *science of wealth* in its three major phases, viz, production, distribution and consumption. Development of economic process influences the social life of man, and the reverse is also true. Thus, the relation between Sociology and Economics is intimate. Some Social Scientists like **Max Weber, Pareto** and **Schumpeter** have interpreted Economics as an aspect of social change. According to them, the study of economics would be misleading and incomplete in the absence of its social setting. Moreover since the economic system is a part of the social structure, the study of economics cannot be undertaken except as the function of Sociology. In sharp contrast to this view, there are other eminent Social Scientists like **Marx** and **Veblen** who believe that the social reality in the last resort is reducible to the interplay of economic forces. These Scientists tend to regard Sociology as an aspect of Economics. Economics studies creation, distribution and consumption of wealth. Economists have developed tools for measuring things like prices, supply and demand, rate of inflation, G N P (gross national product) and money supplies. Here is a simple example to illustrate the difference between Sociology and Economics. If workers in a factory go on strike, an Economist would be concerned with reduced productivity, loss of revenue and its effect on the nation's economy. A Sociologist, on the other hand, would investigate the effect of the strike on the workers' family and workers' social behaviour.

1.5.2 Sociology and Political Science

From the historical point of view, the relationship between Sociology and Political Science is very close and the distinction between the two is fairly recent. **Plato's** *Republic* and **Aristotle's** *Politics* were considered to be comprehensive treatises in Political Science, but these are treatises on social thoughts. **Kautilya's** *Arthashastra* is a great treatise on Economics, Political Science and Sociology. The subject matter of Political Science is social groups organised under the sovereignty of the *State*. This may also be studied as a social institution by Sociologists. The reason is that the *State* is a social structure within which other smaller societies like community and family develop and function. Sociology considers the *State* as a mere social institution. Political Science considers the *State* as a supreme regulating power of the community and the source of political law. Traditionally, Political Science has three main aspects :

(1) Philosophical aspect (political theory).

(2) Practical aspect (actual operation of Government).

(3) Descriptive aspect (political behaviour).

Political Science overlaps with Sociology in its emphasis on political behaviour. **Karl Marx** linked political institutions and political behaviour with the economic system (capitalism and social class or workers). The most significant thing about the close link between Sociology and modern Political Science is that most of the changes, which have taken place in political theory in the last few decades, have been along the lines of development suggested by Sociology.

1.5.3 Sociology and History

As in the case of Political Science, Sociology has strong links with History too. The main difference between Sociology and History lies in the fact that History deals with human events in so far as they are correlated in time, while Sociology studies them from the view-point of the social relationships involved. For example, while a Historian describes a war with all the circumstances accompanying it, a Sociologist investigates the impact of the war on the lives of the people and on subsequent developments. History looks at the past to learn what happened, when it happened and why it happened. Sociology also looks at historical events, but within their social contexts, to learn why things happened and to study their social significances. Historians present a narrative of the sequence of events during a certain period. Sociologists, on the other hand, examine historical events to learn how they later influenced social developments. In spite of very close links between Sociology and History, there are some major differences in their approaches and view-points. Some of these differences are:

(1) History is *concrete*, but Sociology is *abstract*. History presents a chronological description of incidents, cultures, etc. Sociology tries to find out the causes of these incidents and to discover the general principles governing them. In case of wars, History merely describes them, but Sociology attempts to study the social processes which lead to wars.

(2) History and Sociology have *different* attitudes. History, in general, studies those incidents which are peculiar or unusual. Sociology, on the other hand, studies these incidents or processes which occur commonly or repeat frequently.

(3) History studies the events which happened in the *past*. In contrast, the focus of Sociology is on the *present* events.

(4) History emphasizes the activities of powerful or prominent *individuals* (such as Kings or other heads of State), but the unit of Sociological study is the *human group*.

1.5.4 Sociology and Anthropology

The term "**anthropology**" has been coined from two Greek words viz., *anthropos* (meaning "man") and *logos* (meaning "the study" or "the science"). This implies that, Anthropology should deal with *man as such*, while Sociology should deal with him in as much as he is a *social being*. In actual practice, however, the scope of Anthropology is considerably restricted. For example, Social (or cultural) Anthropology, to which Sociology is more related, confines itself mostly to the study of culture chiefly among pre-literate people and small communities. In the last fifty years, there has been a tendency to merge Social Anthropology with Sociology and some pioneering works on Sociology have actually been written by Anthropologists. In our own country, the village studies as well as some urban studies in the early 1950s were mainly initiated by Social Anthropologists. In spite of such close affinity between Sociology and Social Anthropology, the distinction between these two branches of social sciences continues. The reason is that their respective points of views on the social reality are somewhat dissimilar and their fields of study do not always coincide. The Social Anthropologists confine themselves mostly to the study of rural and tribal communities, whereas the sphere of study of Sociologists extends from rural to urban communities and to various problems of the society. There is also dissimilarity in the methods of research used by these two disciplines. Sociology has grown out of Philosophy and History, while Anthropology took roots from Biology. Sociology makes use of documents, social surveys, statistics, etc. Social Anthropology resorts mainly to the functional method in which the person conducting the research actually goes to live in the society he intends to study.

1.5.5 Sociology and Psychology

Psychology is the study of functioning of human brain, mental processes and individual behaviour. Psychology is concerned with issues like emotion, perception, thinking, motivation, creativity, mental disorder and personality. Sociology and Psychology overlap in a field of study known as "*Social Psychology*", which deals with human behaviour and the way it is influenced by various social situations. Social Psychologists study such issues as how the individuals behave in a group, how they solve problems and how they reach or fail to reach an agreement. Sociology and Psychology are closely interlinked with each other and they are complementary. The reason is that without understanding the working of human mind, it would be extremely difficult to understand the human behaviour in the society. Since society is fundamentally a mental phenomenon, the relation between Psychology and Sociology becomes obvious. At the same time, the whole of social life is not reducible to psychological phenomena since the social life is also affected by economic, cultural, geographical and many other factors. As a result, the distinction between Sociology and Psychology remains. Sociology studies the behaviour of a *group*, while Psychology studies the behaviour of an *individual* and the mental processes involved.

1.6 ROLE OF A SOCIOLOGIST

1.6.1 Conducting Research

Like all other scientists, the Sociologists are concerned with both collecting and using knowledge. The foremost task of a Sociologist is to discover and organize knowledge about the society by conducting sociological research.

1.6.2 Correcting Social Prejudices

A very important task of the Sociologist is to clear away the social prejudices and misconceptions, which affect much of our social thinking and create social tensions. Sociologists have helped to remove a great deal of prejudice about heredity, race, sex differences and many other aspects of social behaviour.

1.6.3 Making Sociological Predictions

Another important task of a Sociologist is to make sociological predictions based on basic sociological principles and current trends in the society. Sociological predictions are of interest to legislators, administrators and others who are concerned with any aspect of social policy. Sociological predictions can help to estimate the probable effects of a social policy formulated by government or other agencies.

1.6.4 Other Roles

Sociologists are increasingly finding employment today with government departments, non-government organisations, corporations, welfare agencies, hospitals and other large organisations. In addition to conducting sociological research, they are engaged in planning and executing community action programmes and advising on public relations, employee relations and inter-group relations within an organisation. Sociologists engaged for such tasks are usually specialists who have specialized in Social Psychology, Industrial Sociology, Urban or Rural Sociology. In such positions, the Sociologists work as applied scientists.

1.7 SOCIOLOGY AND THE ENGINEER

Today the professions like engineering, medicines, law and agriculture are increasingly urging more social science course contents in their curriculum. Why should an engineering student learn Social Sciences in general and Sociology in particular? A brief answer to this question is that an understanding of people and society is necessary for success in engineering or in any other profession. The engineers or other professionals are members of a society, residents in a community, participants in many groups and carriers of the culture to the next generation. Study of Sociology helps them to fill their varied roles with greater insight.

There are at least *four* reasons for teaching Sociology and other social sciences (such as Industrial Psychology) to engineers. The most laudable reason was given by **C.P. Snow** while pleading for bridging the gulf between the two cultures – the scientific/technological culture and the literary/humanistic culture. His major concern was that each of these two major cultures in a modern society has a distinct mode of thought, divergent levels of understanding and different languages, while each culture has common attitudes, common standards and patterns of behaviour, and common assumptions. As a result, scholars cannot communicate across the narrow confines of their cultures. This gulf between the two cultures has resulted from over-emphasis on narrow specialization and it can be bridged by exposing the students to both the cultures. The second reason for teaching social sciences to engineering students was given by the **UNESCO** at its twelfth session held in December, 1962. It was recommended that in addition to courses in their area of specialization, technicians, engineers and technologists should study the social and economic aspects of their respective fields in order that most harmonious and effective use could be made of the human and material resources available to them. The third reason for exposing engineering students to social sciences is due to the **Kothari Commission Report** (1964 – 1966). It states, "Technology itself is evolving so rapidly that a student who receives only a narrow and specialized training, to the exclusion of general education in sciences and humanities, will quickly find his skills obsolescent and lacking an adequate base for the complexity of the demands of the modern world. Therefore, all technical education should contain an appropriate element of general education". The fourth reason for social science courses in the engineering curriculum has been given by **Chavarria – Aguilar**, who formulated the Humanities and Social Science (H.S.S.) curriculum for I.I.T, Kanpur in 1962. According to him, an engineer has to assume responsibility and leadership and hence, he must study a good deal of H.S.S.. To that end, such a curriculum must seek to open windows in the narrow house of speciality in order to let light in from other sources; it must seek to bring the individual student to a world beyond his professional interests; it must seek to keep him informed and alert to his environment and to his role in it while providing him with the basic intellectual tools for coping with it. And, perhaps most important, it must seek at the same time to create incentives, to implant in the individual the willingness, the eagerness, in fact, to assume intelligently and effectively the various and increasingly complex roles of modern social man .

MODEL QUESTIONS

(Essay/Long Type)

1.1 Define Sociology in detail.

1.2 What is the scope of Sociology ? Explain in detail.

1.3 What are the major fields of Sociology ?

1.4 Differentiate Sociology from Economics, Political Science, History, Anthropology and Psychology.

1.5 Write briefly the role of a Sociologist.

1.6 Why should a student of Engineering study Sociology ? Explain.

(Objective/Short Type)

1.7 How was the word "Sociology" coined ?

1.8 Name the two famous Sociologists.

1.9 Name both the schools of thoughts regarding the subject matter and scope of sociology.

1.10 What is Social Morphology ?

1.11 Name the two major fields of Sociology.

1.12 What is the main difference between Sociology and History ?

1.13 Define Social Pathology.

1.14 Define Economics.

1.15 Define Anthropology.

1.16 Some of the major fields of Sociology are :

    (a) Demography   (b) Criminology

    (c) Industrial Psychology   (d) All the three.

1.17 Sociology has intimate relationship with:

    (a) Economics   (b) Political Science

    (c) Social Psychology   (d) All the three.

1.18 The comprehensive treatises in Political Science are:

    (a) Plato's Republic   (b) Aristotle's Politics

    (c) Both   (d) Neither.

1.19 Sociologists can be called:

    (a) Engineers   (b) Technologists

    (c) Only (a)   (d) Both.

1.20 Match the following:

    (A) Sociologist   (a) Kautilya

    (B) Political Science   (b) Max Weber

(C) Economics

(D) Philosopher

(c) Auguste Comte

(d) Karl Marx

1.21 Write true or false:

History, Economics and Psychology can be called Social Sciences.

1.22 Name the odd ones:

(a) C.P. Snow

(b) Chavarria-Aguilar

(c) D.S.Kothari

(d) Aristotle.

1.23 Fill up the blanks:

Sociology studies four main problems:

(a) Social Morphology

(b) _____

(c) Social Control

(d) _____

Societal Structures

2.1 INTRODUCTION

Every field of study makes use of certain "technical terms" to which that field attaches special meanings. The reason is that we need basic, clear and meaningful concepts for a scientific discussion. For example, how can one explain the working of an electric motor to a person who has no concept of electric current ? Similarly, a clear idea of basic sociological concepts is necessary in order to understand various social phenomena. Most of the basic concepts in Sociology are expressed in words which also have a popular meaning. The term "order" has a meaning in Sociology, which is quite different from that at a restaurant table. The same is the case with every science, which appropriates some common words and makes them into scientific concepts by giving them special meanings. In this chapter, we shall discuss the basic sociological concepts and their applications to contemporary Indian society.

2.2 SOCIETY AND SOCIOLOGICAL CONCEPTS

In common parlance, the word "society" is often used to refer to a formal association of persons interested in a common cause, (the Indian Chemical Society or the Institution of Engineers (India)). To a Sociologist, however, a society is that group within which the members share a common life. In identifying a society, three characteristics have to be satisfied:

(1) There should be an aggregate of individuals.
(2) These individuals should possess a sense of belonging and co-operation.
(3) Their association should be more or less permanent.

The term "social organisation" is used to describe the vast, complex network of human behaviour within a society. It is obvious that social organisation is a dynamic concept since the patterns of human relationships are constantly changing, although they are more or less regular and predictable. Thus a social organisation can be considered as a condition as well as a process.

A term with a more restricted scope is "social process", which comprises the dynamic elements of a society. As individuals and groups come together and strive for goals and manipulate one another to achieve the things with social values, interactions take place among them. Before an interaction could take place, there should be communication between two or more persons. In a society, certain forms of interactions occur again and again with uniformity and regularity. Such interactions are designated as social processes. There are two types of social processes:

(1) solidary and (2) antagonistic. Co-operation, accommodation and assimilation are classified as solidary processes. On the other hand, competition and conflict belong to antagonistic processes.

2.3 SOCIAL STRATIFICATION

2.3.1 What is Social Stratification?

In any society, each individual occupies several *statuses*. The total number of such *statuses* a person holds makes up what is termed as his or her "station'. Within each society, there are many persons who occupy roughly the same station and thus, they divide the population into horizontal layers or strata. This phenomenon is known as "social stratification". Those in the upper strata of a society enjoy more prestige, honour, wealth and authority than persons in the lower strata of the society. Generally speaking, *social stratification* refers to the ranking of social positions in terms of amounts of certain desirable things associated with these positions. The important and desirable things in most societies are related to wealth, prestige and power. *Wealth*, may be defined as an ability to produce income and/or inherited valuables such as real estate. *Prestige* refers to the style of life and the respect one commands in the society. *Power* refers to the ability of the individual to control or dominate the course of events in a social life. Social stratification, hence, involves inequality in a society. The higher the rank of a given position, the more desirables one can get by holding that position. It may be noted here that the phenomenon of stratification occurs in *all* groups from families to entire societies. In the stratified structure of a society, it is generally the social position of a person's occupation that is ranked. For example, a father may have the highest rank within his own family, but his rank in the larger society usually depends upon his occupation. A person's educational opportunities, physical comforts, influence over others and the amount of respect received from others are directly associated with that person's position in the *stratification* system. Social problems like the presence of poverty in the midst of plenty and the patterns of conflict among different ethnic groups are essentially problems of *social stratification*. This implies that, as a general rule, poverty in a society does not exist because of the lack of material wealth, but because of the highly unequal distribution of wealth.

2.3.2 Why Does Stratification Exist?

Social stratification appears to be inevitable even in social groups that deliberately attempt to eliminate it. Why does stratification always develop?. Why are wealth, power and prestige distributed so unevenly even in an affluent society? There are two major theories to answer such questions: the *functionalist theory* and the *conflict theory*.

The functionalist theory of *social stratification* was proposed by the Sociologists, Kingsley Davis and Wilbert Moore. According to this theory, *social stratification* occurs because the inequality in the distribution of desirables is a device or means for assuring that the most important positions in the society are filled by the most qualified individuals and that the people in these positions perform their assigned tasks more competently and sincerely. The functionalist theory of stratification assumes that, in any society, some tasks are more important and essential than others and that they require more talent, training, or preparation. Desirables like wealth, power and prestige are the social rewards that motivate certain talented individuals to undergo the required training and preparation and to make the accompanying sacrifices and, therefore, such individuals must be amply rewarded with these desirables. Getting the most qualified people into the most important positions is only one aspect of the use of rewards to motivate people in a stratified society. Once, persons have prepared themselves to assume important positions, they must be motivated to do the tasks associated with such positions competently. Greater rewards often go to those who do their work better than others occupying similar positions. According to the functionalist theory of stratification, social inequality is necessary or "functional" for a society. There is an entirely different explanation for the existence of stratification in a society, which is known as the *conflict theory*. This theory contends that social inequality is due to some people having the capacity and willingness to exploit others. They hold a monopoly over the desirables of society and use their monopoly to dominate over others. Conflict theorists believe that societies are run by power rather than by consent. According to the conflict theory of social stratification, the persons, who have certain advantages, use those advantages to gain more advantages for themselves. Those with wealth, power and prestige are able to maintain and perhaps even increase their share of desirable things in the society.

2.3.3 Stratification in Rural India

As an example of the concept of *social stratification* to contemporary society, we shall examine the stratification in rural India since about 70% of our population live in more than half a million of villages. The social organisation of a typical Indian village is a complex frame work, built around the traditional caste structure of the Hindu social system. In fact, even the non-Hindu groups in villages are influenced by the stratification based on the caste system and often acquire a quasi-caste status in accordance with their social, economic and political standing in the village. In central Indian villages, the *Gonds* (a tribal community) acquire the social status in the middle rung of the hierarchy. On the other hand, the *mahars* (a scheduled caste) become Buddhists in order to remove the stigma of a lower social status, but they still remain at the lowest rung of the society. In villages where the scheduled castes are converted to Christianity, their position in the social hierarchy still remains the same and they still occupy the lowest position. Villagers regard the castes as higher or lower than one another in precedence and esteem, there by forming a hierarchy of castes. However, there is no unanimity among the villagers about ranking of castes in a village. Assignment of "rank" to a particular caste may vary from village to village. Some of the criteria entering into the ranking of castes in villages are dietary customs, occupation, details of ritual behaviour, wealth and the interaction of castes with regard to giving and taking of food and water.

2.4 CASTE

2.4.1 What Is a Caste?

A caste may be defined as an *endogamous* (allowing its members to marry within the caste only) and *hereditary* subdivision of an ethnic unit occupying a position of superior or inferior rank (or social esteem) in comparison with other such subdivisions. A caste resembles the *clan* in being a social unit within a larger political or cultural whole, and also in being marriage-regulating and therefore, hereditary from the point of view of the individual. However, a caste differs from the clan in being endogamous, whereas the clan is *exogamous* (allowing its members to marry outside the clan also). Another major difference between a caste and a clan is that the castes, by their very nature, are *ranked* (or rated), whereas clans are essentially *equivalent* within the society, of which they are constituent units. The caste and the clan may be described respectively as horizontal and vertical divisions in a society.

Castes may be considered as a special form of social *classes,* which are present in almost every society. Social classes are the soil from which caste systems have independently grown up at various times and in various places. These caste systems have an outward or analogous similarity, but have often been without historical connection. The similarities between various caste systems of the world are merely conceptual, and not genetic. No single factor (or a set of factors) is to be regarded as the universal cause of the caste system. Conquest, race difference, religious practice, economic development and many other factors have contributed to the growth of caste systems in many parts of the world. The phenomenon of castes is not uniform, but shows wide differences in various societies. There are some societies (the Indian society), which are consciously caste organised.

2.4.2 Caste System in India

Although castes have existed in some forms in various societies of the world, it is in India that caste and caste system has its full development. Caste has become an intricate system in India. The Indian caste system forms the most integrated as well as the most self-conscious system that has grown up anywhere in the world. However, the caste system in India appears to be less than three thousand years old. The *Vedas*, believed to have been composed between 1000 and 1500 B.C., do not mention substantially the caste system. By the time of Buddha (around the fifth century B.C), however, the caste system was fairly prevalent.

According to the Hindu theory, four major castes were instituted at the very beginning of human civilization. These four castes are : *Brahman* (priests), *Kshatriya* (warriors or nobles), *Vaishya* (cultivators or traders), and *Shudra* (labourers or serfs). The multiplicity of castes originated from marriages among the four original castes.

The census of India has recorded over eight hundred castes and sub-castes. If minute or wholly localised castes and sub-castes are considered, the total number becomes nearly five thousands. These include the occupational groups and also various tribes, races and sects. The Parsees (the fire-worshipping refugees from ancient Persia) are treated as a caste; and so are the Todas and many other primitive hill tribes. The same is true of fishermen, scavengers, jewelers and, in some places, even Christian converts. There are some social groups like Anglo-Indians who are self-sufficient, whose racial or cultural origin is foreign and who are not really interested in constituting a caste; but even such groups are forced into the mold of caste system by Hindu opinion.

New castes (or rather sub-castes) are still forming today by fission. For example, members of the same caste may be forced by economic conditions to follow somewhat different trades in two adjacent areas. If one of these two trades ranks a little higher in the social hierarchy, two sub-castes arise and one of them begins to look down on the other as a poor relation. In addition, a part of a group migrating into a new region may find its occupation more lucrative and more highly rated than the earlier occupation and before long, a new caste arises. Every new religious cult carries with it the potentiality of a new caste (the Kabirpanthis or the Radhasoamis), as does every new conquest made by a population group rather than by an army.

2.4.3 Origin of Caste System

The history of the caste system in India is extremely complex and its causes are manifold. Any attempt to explain the caste system on the basis of a single factor may appear to be attractive, but is foredoomed to incompleteness and failure. One attempt can be based on the *racial explanation* of the origin of Hindu caste system. According to this theory, higher castes mostly consist of fair-skinned Aryans, while the members of most of the lower castes belong to the dark-skinned Dravidian race. This explanation, however, is too simplistic. It is true that one of the Hindu words for caste is *Varna*, which literally means color. Based on anthropometric records (records related to various physical characteristics of human beings), higher castes have lighter skins and narrower noses (the basic features of Aryans). On the other hand, social rankings of various castes and their physical indices do not agree *uniformly*, even in the same area. Members of high castes in south India are often broader-nosed than those of the low castes in north India. It is quite likely that race consciousness might have given the first impetus to the caste system in India; but economic status is now a much more potent factor than race consciousness for the creation of castes and sub-castes.

Another major factor responsible for the origin of the caste system is religion, which has deeply influenced Hindu castes. Members of higher castes are believed to have earned the privilege of being born in higher castes by meritorious deeds in their past lives. Since the members of higher castes represent greater merit, more duty is incumbent upon them and, consequently, they are subject to numerous restrictions regarding food, sexuality, cleanliness, conduct, etc., from which lower castes are exempted. A Brahman may eat only vegetarian food and that too, prepared by another Brahman; but the lowest caste may eat anything including pork. The Hindus, in general, do not feel caste as a burden as the individualistic people from western countries might feel. To a Hindu, the caste system seems both natural and desirable, and its deliberate breach is considered unnatural, perverse and unforgivable. Whatever their caste, most Hindus are proud of it in the same way that the westerners are of their nationality. The caste gives a Hindu a sense of solidarity.

2.4.4 Recent Changes in the Caste System

During last five decades, the caste system in India has undergone profound modifications. A glaring example of this change is that the Indian Constitution has forbidden *untouchability,* the most undesirable feature of the Hindu caste system in any form. Now, the temples and the wells have been thrown open to the depressed classes of the society. No discrimination can be observed now in public halls, recreation centres, schools, colleges and universities. The scheduled castes, the scheduled tribes and the other backward classes have been given many safeguards like reservation of seats in the parliament, state assemblies, village *panchayats,* educational institutions, government jobs, and scholarships for education. In addition to the role of national leaders like Mahatma Gandhi and Dr. B. R. Ambedkar, the spread of education and communications as well as the multiplication of personal contacts have been instrumental in changing the traditional attitudes and values related to the caste system. Many old types of occupations are disappearing now and many more are being created. In the new occupational setting, Brahmans, Kshatriyas, Vaishyas and Shudras meet freely and work together. The new economic and occupational opportunities are open to the members of all castes. This has given rise to competition among the members of the same caste. As a result, the caste that was traditionally the horizontal layer in the hierarchical order of the society has now taken a vertical movement.

The main factors responsible for bringing about the flexibility in the rigid caste system in the Indian society may be summarized as follows:

(1) The process of industrialization and urbanization.
(2) More frequent contacts between the rural and urban population.
(3) Universal education.
(4) Social legislation leading to amelioration of the status of lower classes.
(5) Spread of transport and communication systems.
(6) Changes in the occupational structure.
(7) Social reform movement.
(8) Introduction of democratic institutions at various levels.

2.5 CLASS

2.5.1 Concept of Class

The concept of class is concerned with the social differentiation of groups. In ancient times, this differentiation was related to the status or rank in a society and not to the class. Membership in a particular station of the society was bound up with privileges in the realm of public and private laws. Differentiation according to status was fixed. A man was born into a certain social station, law and custom determined his membership in it and a change of status was practically out of question. In contrast, one may speak of *classes* as existing when a transition from one social group to another is easily possible for the individual, either personally or in the course of generations, and when such a transition takes place rather frequently. These two principles of social organisation (status and class) are not easily distinguishable since historically they are found existing side by side and often overlapping.

In our daily speech, the term "class" is used in many different senses. We talk of upper, middle and lower classes; of propertied and non-propertied classes; of productive and unproductive classes, of educated and uneducated classes, and so on. In this general sense, the term "class" is practically meaningless since it only says that a group of people have certain characteristics in common. To give this term a more precise meaning and thus to improve its scientific utility, it must denote a social group where the characteristics held in common are perfectly definite and already determined. Occupational groups must fall outside the category. The general concept of the work "class" today has reference not to occupational groups (where the common characteristic possessed by its members is external and superficial), but to *social* classes. Thus, the concept of *class* is intended to analyse the social structure and the social stratification of the population.

There are two schools of thoughts on the question of what is essential in the concept of social classes - the first, selecting the *objective factors* at the basis of social class and the second, selecting the *subjective factors*. In the first school, some authors regard the ownership or non-ownership of the instruments of production as basic to the concept of class, while others lay more stress upon the general standard of living. Additional objective factors have also been used as the basis of a class. The eminent Sociologist, Max Weber has built the concept of a class upon three factors :

(1) possession of economic means (2) external standard of living and (3) cultural and recreational possibilities. According to the second school of thought consisting of subjectivists, social classes are groups whose sources of income are similar and whose economic interests coincide. In this formulation of the concept of social class, the subjective factor lies in common economic interest and outlook. The renowned sociologist Sombart, who has stressed the subjective aspect most sharply, regards a social class as a group which by its way of thinking stands for a particular system of economic organisation. Common economic interest and common ideology are considered important factors in the subjective point of view. Some other sociologists regard the degree of esteem in which a group is held as essential

to the concept of class. These theorists make classes essentially a gradation of ranks based on social prestige. They classify the gradation of ranks into social strata on the principle of *"the subordination of the intellectually inferior to the intellectually superior"*.

2.5.2 Historical Evolution of Class

In tracing the historical evolution, it is necessary to include all the situations where gradation of ranks occurs, associated with differences of social position and variation in both outer and inner conditions of life. Another premise that is necessary in order to distinguish class from status is that *mobility* from one group to another is possible and that it occurs frequently. Using such a criterion, one finds that even in primitive societies, group structures occurred which had unmistakably the character of classes. Where a community is not a closed society with a caste system, certain individual members of the community are often set apart from the rest because of acquisition of wealth or display of unusual craftsmanship. **Fahlbeck** and other sociologists point out that the transition from a society of status to that of class occurred in Greece during the seventh and the sixth centuries B. C. and in Rome, somewhat later. The class conflict in Greece arose from the opposition of peasantry, heavily in debt, to the landed aristocracy. A similar development occurred in Rome. A social differentiation on the basis of status, by which political power was concentrated among the land owners was gradually broken down and the way was opened for differentiation into classes. When the transition became complete, the development of Roman social organisation continued in the direction of sharpening the outlines of class structure. After the subjugation of Italy, the inequalities in wealth were accentuated and the newly emerging moneyed classes won increasing power and obtained a position. Largely because of the wars and the competition of grain from other countries, the peasant class disintegrated. Deprived of their land and livelihood, the peasants thronged to the metropolis (Rome) where they constituted an enormous proletariat and led a meagre existence on public dotes. This led to the emergence of definite property classes, with the sharpest contrast in the distribution of wealth. With the economic collapse of the Roman empire, the peasants were hereditarily bound to the soil and Italy reverted to the old system of occupational status.

In the middle ages, the feudal system represented a social organisation based on status. With the increasing importance of production for the market and trade and with the coming of economy based on money, factors in addition to those related to descent began to affect social differentiation. Gradation arose among the *free* people and those who were *less free*. Neither ranks nor classes remained permanent (or unchanged), but disintegrated into subgroups. Later on, gradation of this sort began to arise among the *unfree* people as well. At the other end of the scale, with the strengthening of princely power, there emerged a new aristocracy of *office* besides the old aristocracy of birth. This new aristocracy consisted of the ministers in the service of the prince. With the growth of cities and trade and the development of a money economy, another occupational class- *the burghers*– took its place with the aristocracy and the peasants. Office and occupation slowly began to determine social position and members of the most diverse ranks by birth found it possible to move to higher social strata. While the new classes were occupational, they remained quite rigid at the outset, but wealth and vocation kept continually displacing the facts of birth and descent. Although one was born to a certain station in society, it became more frequently possible to move into another. From the sixteenth century, economic and intellectual developments have been leading to many changes in this system of social organisation. While this evolution of class was essentially similar in all European states, the manner and time in which it took place differed significantly, particularly in England and France. The class organisation evolved earlier in Italy than in other European countries. The wealthy merchant class had attained an influential position in the British Parliament by the end of the seventeenth century. Around the same time in France, numerous burghers had been elevated to the nobility and after 1715, they could acquire the estates of the nobility.

It was only at the beginning of the nineteenth century, that the changes in the distribution of wealth attained so great an influence upon social stratification and social mobility that they transformed the very nature of the class structure. With the growth of capitalism and large scale industry, the economic element (possession of property) played a greater role than ever before in the determination of class membership. The purely social factors yielded almost entirely to the economic factors, and this led to a huge increase in the number of people whose lot it was to always belong to the wage-earning class because of their lack of property. With the advance of socialism, there came to the fore an ideology deriving from the particular situation in which the working class found itself. It was becoming increasingly obvious that the membership in a certain class was determined by the distribution of wealth and income. In this period, when education was increasing and when growth of cities made social contrasts stand out more strikingly, the economic plight of a class could have a very deep effect on the consciousness of the group concerned. In the course of historical development, new criteria for the essential nature of classes have arisen to supplement the old ones. The emergence of such a criterion as the subjective one of a common economic ideology had to wait for the developments in the nineteenth century. Before that, the workers as a class were too under developed to have a common consciousness of their plight and their interests. While there have been instances of class conflict in history, the statement of Karl Marx that "the history of all society up to now is the history of class struggle" should be regarded as too sweeping. The class struggle, as motivated by a class conscious opposition to the existing capitalistic system, is a fairly modern phenomenon.

2.5.3 Formation of Classes

In the last century, this problem was the subject of controversy between two eminent social scientists, Gustav Schmoller and Karl Bücher. According to Schmoller, classes were the result of the division of labour. He asserted that different occupations had certain specific physical and psychological effects upon individuals which were transmitted by the natural selection of variations through heredity. Bücher, on the other hand, asserted a causal relationship running in just the *opposite* direction. He claimed that the differences in wealth and income were the *cause* rather than the result of the division of labour. It may be noted here that Bücher's view seems to be correct so far as the formation of *modern* classes is concerned, though the division of labour might have played a role, as suggested by Schmoller, in the *original* formation of social classes. While the major forces in class formation have been social in early times and mainly economic in more recent times, certain qualifications must be added. In the nineteenth century, there were attempts to explain the formation of social classes as a *socio-biological effect*. According to Sir Francis Galton, the main proponent of this theory in England, social classes were regarded as the outgrowth of the natural inequality and native differences of men in the struggle for existence. The supporters of this theory see the social elevation of individuals and the resulting formation of classes as a consequence of extraordinary ability and achievement. Such a view can have *complete* validity only where the competition occurs among individuals who are at an *equal* level of advantage. Under modern economic conditions, this is obviously an untenable postulate. On the other hand, there is some value in this theory as native endowments and achievements play a significant role in the formation of classes.

Unfortunately, statistical science all over the world has not paid enough attention to the manner in which the formation of class takes place, although relevant data could easily have been obtained. In the USA, the opportunity for a worker to rise higher in his own class and even out of it into another class is greater than in Europe and most other countries, and his social position carries more respect.

The present class structure of society has been attacked from two directions. The socialists, on one hand, have a *classless* society as their idea; but it is doubtful whether this goal can ever be completely achieved. There are other social thinkers, on the other hand, who view modern social classes as constituting a principle of social division and they are deeply concerned with considerations of social equality. In the last few decades the conviction has grown that, while the earlier systems of status and rank arose spontaneously out of a common way of life and subordinated their interests to it, classes are self-seeking and disintegrating groups in a society. Many in Europe, therefore, look forward to displacement of the class structure by a return to the division of the population according to ranks determined by occupation.

2.6 CULTURAL HERITAGE

2.6.1 What Is Culture ?

Man varies in two respects: (1) Physical Form and (2) Social Heritage or Culture. The science of Physical Anthropology has succeeded in classifying various branches of mankind according to physical structure and physiological characteristics, but man varies also in an entirely different aspect. Consider, for example, a tribal infant transported from his native village in Jharkhand to Japan. Such an infant, when grows up to manhood, would differ profoundly from what he would have been if he was reared in his native village surrounded by hilly forests. Now he will have a different social heritage, a different language and different habits, ideas and beliefs. He would be incorporated into a different social organisation and cultural setting. This social heritage is the key concept of Cultural Anthropology which is another branch of the comparative study of man. The social heritage of man is usually known as Culture in modern Anthropology and Social Sciences.The word "culture" is often used synonymously with "civilisation". The concept of *Culture* in its broader sense comprises inherited artifacts, goods, technical processes, ideas, habits and values. Social organisation cannot be really understood except as a part of culture.

In order to live, man continually alters his surroundings. He creates an artificial and secondary environment at all points of contact with the outer world. He constructs shelters, procures food by means of weapons and implements, constructs roads and uses various means of transports. Were man to rely exclusively on his anatomical equipment alone, he would soon perish from hunger and exposure to heat, cold, fatal diseases, etc. Even in the most primitive modes of human life, all vital needs like defence, feeding and movement in space are satisfied directly or indirectly by means of artifacts. These material outfits of man (his artifacts, buildings, sailing boats, implements and weapons) are the most obvious and tangible aspects of culture. On the other hand, the material aspect of culture is not a force in itself since *knowledge* is also necessary for production, management and use of artifacts, implements, weapons, etc. The material culture requires a complement consisting of the body of intellectual knowledge, the system of moral, spiritual and economic values, social organisation and language. On the other hand, material culture is an indispensable apparatus for molding of each generation of human beings. The outfit of material culture is a laboratory in which the reflexes, the impulses and the emotional tendencies of the human beings in a particular society are formed. The understanding of cultural heritage is found in the process of its production by succeeding generations and in the way in which it produces in each new generation. Therefore, culture consists of the body of commodities and instruments as well as customs and mental habits, which work directly or indirectly for the satisfaction of human needs.

2.6.2 Culture and Institutions

Culture is a well-organised unity divided into two *fundamental* aspects, which are a body of artifacts and a system of customs. There are, however, further subdivisions or units of culture. The real units of culture which have a considerable degree of permanence, universality and independence are the organised systems of human activities called *institutions*. Every institution is centered around a fundamental human need. In addition, the institution unites a group of people in a co-operative task and has its particular body of doctrine and techniques of craft. The effective satisfaction of primary needs imposes upon every culture is a number of fundamental aspects like institutions for nutrition, institutions for mating and propagation, and organisations for comfort and defence. The organic (or biological) needs of man form the basic imperatives leading to the development of culture.

The cultural mode of satisfaction of the biological needs of the human organism creates new conditions and thus imposes new cultural imperatives. In all cultures (including the most primitive ones), staple food is prepared and taken according to strict rules (within a definite group) and with the observance of prescribed manners and taboos. The food is usually obtained by a more or less complicated and collectively carried out process (such as agriculture, exchange, community distribution, etc.). Man is dependent on the apparatus of agricultural implements, hunting weapons and fishing craft and also, upon orgainsed co-operation and moral values. These are as indispensable to man's satisfaction of his nutritive needs as the raw material of food and the process of its ingestion. Man is so molded that if he were deprived of his economic organisation and his implements, he would starve as effectively as if the substance of his foodstuffs was withdrawn from him. An amalgamation of functions occurs in most human institutions. The household, for example, is not simply a reproductive institution. It is also one of the main nutritive institutions and an economic, legal and often a religious unit. Family is the place where cultural continuity is served through education. The parents, who develop in their offspring economic attitudes, technical dexterities, moral values and social duties through education and training, have also to hand over at a later stage their possessions and their status or office. The domestic relationship, therefore, implies also a system of laws of inheritance and succession.

2.6.3 Culture and Traditions

The primary biological needs of a community (the conditions under which a culture can thrive, develop and continue) are satisfied in an indirect manner, which imposes secondary conditions. The whole body of material culture must be produced, maintained, distributed and used. In every culture, therefore, a system of *traditional* rules or commands is found. These traditions define the activities, usages and values by which food is produced, stored and distributed, goods are manufactured, owned and used, and various tools are prepared and utilised for production. Regulated co-operation exists even in such simple activities as the search for food among the most primitive communities. Co-operation means sacrifice, effort, subordination of private interest, inclination to the joint ends of the community and the existence of social constraint. Life in a community offers various temptations, especially to the impulses of sex and, as a result, a system of prohibitions and restraints is unavoidable. Similarly, economic production provides man with things desirable and valuable, but not unrestrictedly accessible for use and enjoyment to everybody alike. As a result, rules of property, possession and use are developed and enforced.

Rules of conduct must be drilled into each new generation through education and provisions must be made for the continuity of culture through tradition. The first requisite for this is the existence of symbolic signs in which condensed experience can be handed over from one generation to another. Language is the most important of such symbolic signs. A language itself does not contain experience, it is rather a system of sound habits which accompanies the development of cultural experience in every human community and becomes an integral part of this cultural experience.

Tradition generally remains *oral* in primitive cultures. The speech of a primitive tribe is full of set sayings or proverbs, maxims, rules and reflections which carry on the wisdom of one generation to another. Folk tales, folk songs and mythology form other aspects of verbal tradition. In higher cultures, *writing* is added to carry on the spoken tradition. In primitive societies, education hardly required specific institutions. The family, the local community, initiation campus and professional groups or guilds of technical craft were the institutions which corresponded in some of their derived functions to schools in more advanced cultures.

2.6.4 Culture and Knowledge

Inspite of various theories about the non-empirical character of primitive mentality, there is no doubt that as soon as man developed the mastery of environment by the use of implements and as soon as he began to use language, there must also have existed primitive knowledge of an essentially *scientific* character. No culture, however primitive, could have survived if its arts, crafts, weapons and economic pursuits were based merely on mystical and non-empirical concepts. When human culture is approached from the pragmatic or technological side, it is found that primitive man was capable of exact observations, sound generalizations and logical reasonings in all those matters which affected his normal activities and basis for production. Knowledge is an absolute derived necessity of culture. Knowledge is more than merely a means to an end. Its place in culture and its function are slightly different from those of law or education. Systems of knowledge serve to connect various types of behaviour. They carry over the results of past experiences into future enterprises and allow man to coordinate and integrate his activities. Knowledge allows man to carry on the work which his culture makes him to do. Its function is to organize and integrate the indispensable activities of culture. The material embodiment of empirical knowledge consists of arts and crafts, technical processes and rules of craftsmanship. The integrative function of knowledge creates new needs and it imposes new imperatives. Knowledge gives man the possibility of planning ahead and it allows a wide range to his hopes and desires. On the other hand, empirical knowledge is unable to completely control chance, to eliminate accidents and to see the unexpected turn of natural events. As a result, there develops a special type of ritual activity, known collectively as *magic* in the science of Anthropology.

The most hazardous of all human enterprises known to primitive man was sailing. In the preparation of his sailing craft, the savage turned to his science (his empirical knowledge). But adverse winds (or no wind at all), rough weather and reefs were always liable to upset his best plans and most careful preparations. He had to admit that neither his knowledge nor his painstaking efforts was a guarantee of success. Something unaccountable usually entered and upset his anticipations, but this mysterious element appeared to contain some inner logical consistency. Man felt that he could do something to wrestle with this mysterious force and to improve his luck. That is why, there are always systems of superstition and ritual associated with sailing. In many primitive communities, the magic of sailing craft has been highly developed.

Among primitive people who rely on their fields or gardens for their livelihood, agricultural knowledge is invariably well developed. They know the properties of soil, the need for clearing the weeds and fertilizing the field, the proper season for planting and so on. Inspite of these, mishaps do occur. Drought or deluge may destroy the crops altogether, or some insects may diminish them. The dreaded elements of rain and sunshine, blights and pests seem to be controlled by a force beyond human experience and knowledge. As a result, man resorts once more to magic.

2.6.5 Culture and Magic

Experience and logic teach man that within certain limits, knowledge is supreme. Beyond these limits, however, nothing can be done by rationally founded efforts. Yet, man rebels against inaction since he is driven to action by intense desire and strong emotion. In many cases, inaction is not even a possible option. For example, once a native has embarked on a distant voyage or is halfway through the cycle of growing food grains, he tries to make his frail canoe more seaworthy by charms, or to drive away pests by rituals. Such rituals and charms are examples of magic. Magic changes its forms and shifts its ground, and it exists everywhere and in all cultures. In modern western societies, magic is associated with the third cigarette lit by the same match, with broken mirrors, with passing under a ladder, with number 13 and with Friday. In many Indian societies, it is associated with a cat crossing one's path and with someone sneezing when one starts a journey. Of course, these are just minor superstitions among the intelligentsia of modern societies; but much more developed systems of magic also persist among modern urban populations. In the slums of London, black magic is practiced by the classical method of destroying an enemy by ritually destroying his picture. At marriage ceremonies in western societies, good luck for the married couple is obtained by the strict observance of several magical methods like throwing of slippers and spilling of rice. Wherever there is danger, uncertainty, large possibility of accident or a great deal of chance involved, magic crops up even in an entirely modern form of enterprise. The gambler at Monte Carlo (France) or in Las Vegas (USA) develops magical systems to improve luck. Similarly, mascots are used in motoring and modern sailing. Aviation is developing its own superstitions and magic. In all large cities of America, Europe and Asia, magic can be purchased from palmists, astrologers and others who predict the future, give practical advice for lucky conduct and sell ritual apparatus like mascots and talismans to bring better luck.

It may be noted that magic is never used to replace work. In gardening, for example, the digging and clearing of the ground, the strength of the fencing and the quality of support are never scamped because stronger magic has been used over them. The native knows quite well that mechanical construction must be produced by human labour according to strict rules of the craft. He also knows that various processes in the soil can be controlled by human efforts to a certain extent, but not beyond that. It is only this *beyond* (the region out of his direct control) that he tries to influence by magic. Far from being a primitive science, magic is the outgrowth of a clear recognition that science has its limits and that human mind and human skills are impotent beyond those limits.

2.6.6 Culture and Mythology

A strong belief in magic finds its public expression in the running mythology of magical miracles. Thus, mythology is always found in company with magic. The competitive boasting of one community against another and the fame of outstanding magical successes create a tradition which always surrounds famous magicians or famous systems of magic with a halo of supernatural reputation. Myth of magic is believed to be a warrant of its truth and a charter of its claim to validity. This is true not only of magical mythology but also of other types of mythologies. Myth, in general, is not an idle speculation about the origin of things or institutions nor is myth the outcome of the contemplation of nature and interpretation of its laws. The function of myth is neither symbolic nor explanatory. Myth is the statement of an extraordinary event which has established the social order of a tribe or some of its economic pursuits, its arts and crafts or its religious beliefs and ceremonies. Myth is not simply a piece of interesting fiction, kept alive by the literary interest of the story. Myth is a statement of primeval reality which lives in the pursuits and institutions of a community. It justifies the existing order in the community and supplies a pattern of moral values and magical beliefs.

To conclude, myth is not a mere prototype of literature or science, nor is myth a branch of art or history. In fact, a myth fulfills a function closely connected with the nature of tradition and belief, with the continuity of culture and with the human attitude towards the past. The main function of myth is to strengthen the tradition and to endow it with a greater value and prestige. This is achieved by tracing the tradition back to a higher, better, more effective and more supernatural reality of initial events.

2.6.7 Culture and Religion

The place of religion in the scheme of culture may be considered as a complex satisfaction of certain *derived* needs. The existence of strong personal attachment and the fact of death are perhaps the main sources of religious belief. The affirmation that death is not real, that man has a soul and that this soul is immortal, arises out of a deep need to deny personal destruction. It may be noted that this need for immortality is not a psychological instinct. It is determined by culture and by the growth of human sentiments. To the individual who faces death, the belief in immortality and the ritual of last comforts confirm his hope that there is a hereafter and that it may perhaps be better than the present life. The same can be said about the Hindu belief in the transmigration of the soul and reincarnation. The ritual before death confirms the emotional outlook which a dying man needs. After the death, the bereaved are thrown into an emotional chaos, which might become dangerous to them individually and to the community as a whole, were it not for the ritual of last rites of cremation or burial and all the prayer for the departed soul are acts expressing the belief in the continuity of life after death and of communion between the dead and the living. Any survivor who has attended a number of such mortuary ceremonies for his relatives and friends is prepared for his own death. The belief in immortality, which he has lived through ritually and has practiced in the case of his parents or friends, makes him cherish even more firmly the belief in his own future life. Thus the belief in human immortality, which is the foundation of ancestor worship and mortuary rituals, grows out of the constitution of human society.

Religion sanctifies human life and conduct in its ethics. As a result, religion becomes perhaps the most powerful force of social control. In its dogmatics, religion supplies man with strong cohesive forces. Religion grows in every culture since knowledge, which gives foresight, fails to overcome fate. The cultural call for religion although indirect is rooted in the way in which the primary needs of man are satisfied in culture.

2.6.8 Culture and Arts

Art appears to be the most exclusive and, at the same time, most international of all cultural activities. Music is without doubt the purest of all arts since it is least alloyed with intellectual or technical matters. Decorative arts consist of the ornamentation of the body, of various colours and shapes of clothing, of the painting and carving of objects and of drawing or painting. Plastic art, sculpture, architecture, wood or stone structures, etc., are fashioned according to certain aesthetic criteria. Poetry and dramatic arts are also present in almost all cultures.

All artistic manifestations operate through the direct action of sense impressions. Thus the tone of human voice or of vibrating chords or membranes, the words of human language, colours, body movements, etc., are sensations and sense impressions. These produce a specific emotional appeal, which is the primary element of art since it constitutes the essence of aesthetic appeal. At the lowest scale of artistic enjoyment are the effects of chemical sense impressions (taste and scent), which also produce a limited amount of aesthetic appeal. The main function of art is to satisfy the deep craving of the human organism for combinations of blended sense impressions. Art becomes associated also with other cultural activities and thus develops a series of secondary functions. It is a powerful element in the development of crafts.

2.6.9 Culture and Human Needs

Culture transforms individual human beings into organized groups and gives these groups an indefinite continuity. Culture deeply modifies the basic human endowment. In doing this, it not only bestows blessings, but also imposes certain obligations and demands for the surrender of many personal liberties for the sake of common welfare. The individual has to submit to law and order, he has to learn and obey tradition, and he has to adapt his nervous system to a variety of habits. The individual works and produces things for others to consume, while he is always dependent upon the toils of others. Although culture is primarily born out of the satisfaction of biological needs, its very nature makes man into something essentially different from a mere animal organism.

2.7 OCCUPATION

2.7.1 Concept of Occupation

In any modern society, with its characteristic division of labour and specialization of functions, the overwhelming majority of people are engaged in a specific and relatively continuous activity in order to earn their livelihood and to maintain a definite social status. In the literature of Social Science, this activity is termed as "occupation". According to the Sociologist Simmel, the concept of occupation is fundamental to the true society. It is no surprise, therefore, that the modern society is organized on an occupational basis. Occupation distinguishes one man from another and occupation offers every individual a feeling of independence. At the same time, occupation unites men of different origins and background and serves as a link binding them to the society. The term "occupation" must cover three different aspects :

(1) Technological – the specific mental or manual operations involved in the execution of the occupational work.
(2) Economic – the income yield of an occupation which serves to provide a livelihood.
(3) Social – the prestige attached to a person or a group by virtue of occupation.

In accordance with the current popular and sociological usage, the term *occupation* may be defined as that specific activity with a market value which an individual continually pursues for the purpose of obtaining a steady flow of income. This activity also determines the social position of the individual. Occupation is closely linked up with the present socio-economic system in which production is for sale and not exclusively for personal use. Thus occupational work must be available to purchase goods and services in the open market. From the individual's point of view, occupation is the specific activity by which a living is earned. From the point of view of a Social Scientist, occupation refers to an individual's innumerable activities. Occupational work is the individual's contribution to the national economy, and the sum total of occupied persons constitutes the labour force of a nation. The relationship between the individual, his occupation and the society are subject to certain qualifications. For example, not everyone has an occupation. Apart from children and those still under training for their vocation at the lower end of the age scale and the sick and aged at the upper end (all of whom must be supported by the working population), there are in any modern society persons of independent means (landlords or pensioners) who share social products without pursuing any real occupation. In modern times, continuous unemployment has created a new group of people who live without working. In the second place, an individual may simultaneously engage in several trades, one of which will be recognized as his main occupation as it determines his social position and yields the major portion of his income. Occupation, moreover, is not a rigidly limited hereditary activity in modern society, and changes of occupation occur quite frequently in industrial societies.

Occupational lines are shifting, the personnel of any specific occupation is mobile, but the proportions between the larger occupational strata remain fairly stable over a long period. This phenomenon may be explained on the basis of two forces at work – one making for change and the other tending to maintain the balance. With regard to the social division of labour, which lies at the basis of occupational differentiation, these two forces may be identified as the rational adaptation to the demands of the economic system and the traditional adherence to the institutions of the past. These two forces are combined in varying proportions in different societies.

2.7.2 History of Occupation

Throughout the greater portion of history, society in all its manifestations has been shaped primarily by *religious* authority. The force of tradition in modern society is perceptible in the allocation of high prestige value to certain occupations which can scarcely be related to their social utility. A rough correspondence of prestige to income is found only in the occupations related to business. In other fields, especially in liberal professions, the social position is largely independent of the lucrativeness of activity, representing the continuation of the values of the past. Since occupation is not simply an economic or a technological fact, occupation evolution is neither a mere part of economic history nor an aspect of technological progress. It is rather an important constituent of general social history. In occupational development, industry has always been the dynamic factor because of the possibilities for division of labour. The oldest and the earliest distinct occupation in all societies has been the "spiritual" professions like those of magician, sooth-sayer, prophet, medicine man and singer. Such professions possess a charismatic monopoly. Social division of labour began with the appearance of the so-called wage work, the stage in which the worker owned the tools and the customer provided the raw materials. Like the early "spiritual" professional, the craftsman assumed the performance of certain public functions.

The period of wage work continued in Europe until the fourteenth century. By that time, money economy in the cities had made an appreciable headway. In the new system, the worker owned not only the tools but also the raw materials and, in some cases, even the workshop. At this stage, crafts became organized into guilds. These guilds were public corporations which furthered the economic interests of their members and at the same time, assumed the responsibility for providing needed goods of assured quality at reasonable prices. On the economic side, there was a vertical split in industrial production, and various phases of the manufacturing process were separated into independent industrial branches. This necessitated mutual exchange between specialized economies, which now produced for the outside consumption rather than for the satisfaction of their own needs. Finally, the political significance of the guild system developed into a system of industrial self-government. The regulated economy of the guild society disintegrated with the growth of a centralised state organisation, unification of large economic areas and radical changes in the techniques of production and distribution, which took place during the seventeenth and the eighteenth centuries. Mechanisation of industry resulted in regrouping of occupations and eventually, in the creation of a new occupational order. Whereas the division of labour in the handicraft system multiplied independent industrial units, the division of labour in the factory created a new and permanent class of wage workers. The industrial society was divided into two antagonistic classes: the capitalist entrepreneurs and the proletariat.

In the factory, the division of labour involved a technically rational breaking down of productive operations into more or less simple manipulative acts entrusted to separate workers, differentiating skilled work from the unskilled. Factories were thus based upon permanent categories of skilled and unskilled labour. Factory workers never mastered all the stages of production. They were aided and directed by a new class of higher industrial officials trained in technical and managerial aspects.

Thus, the true proletariat began to emerge since the free labour now owned neither tools and raw materials, nor workshop. He had to submit to a discipline imposed by a technical process which he did not even understand. The new industrial system and occupational differentiation led also to a change in the relation between occupation as a social grouping and the society as a whole. Corporative organisations of the handicraft era were incompatible with the progressive outlook and competitive spirit of the new system. These organisations, therefore, disintegrated under the pressure of territorial and national governments. Occupation now ceased to be an office, a duty recognised and enforced by authority. Instead of that, occupation became merely a social function, freely assumed by each individual after balancing his personal inclination against economic necessity.

2.7.3 Current Trends

Although the formal freedom of occupation dispensed with many obstacles to the attainment of this ideal, a great many factors still operate to prevent a perfect correlation between the individuals' capacity and talents, and their occupational utilisation. On one hand, improvements in transportation, communication and organisation of the labour market favour greater geographical and occupational mobility of labour. On the other hand, immigrations barriers check the inflow of labour from outside. Similarly, social insurance provisions bind the worker to his old residence, occupation, or even his job. In a stratified society, moreover, the class factor has a significant influence upon its occupational structure. It is a well known fact that, among individuals, the distribution of physical capacities and mental faculties is very uneven. Moreover, the discovery, cultivation and development of such physical and mental gifts are not equally available to all. Since vocational training is usually a costly and a long drawnout process, it creates an obstacle for admission to an occupation. It is a paradox of modern society, in which a considerable amount of occupational mobility is virtually indispensable, that most occupations require some special training and that a change of occupation usually requires some degrees of individual sacrifice and social waste. It is not surprising, therefore, that workers in a particular occupation develop a spirit of exclusiveness and tend to regard their job as a vested right to be passed on to new comers only under certain specified conditions. The conditions may be a long apprenticeship and a high initiation fee in the skilled trades, a special social background in some liberal professions and government services, or simply contact with the right people, so often invaluable in securing higher paid white collar jobs.

Racial differences and national attitudes also disturb normal occupational distribution. In a few countries, work is regarded as a necessary evil and a tedious preliminary to a life of leisure supported by a fixed income from profitable investments. In many other countries, the inactive investor, landlord or rentier is held in contempt. Similarly, aliens tend to engage in pioneering occupations and thus play an important role in revolutionizing economic practices and in building up new countries. The importance of national factor is clearly illustrated by the occupational distribution of different ethnic groups in the USA, where certain employments attract while others repulse a particular national or racial group.

With regard to occupational employment of women, the assimilative capacity of modern industrial society is far greater than that of the earlier non-industrial society. The gender barrier to occupation, which was one of the oldest and most impregnable barriers, has almost disappeared in modern society, with the result that women are now allowed to join even the armed forces. Of course, there is considerable difference among various occupations in their actual accessibility to women. Moreover, the proportion of gainfully employed women in the total labour force varies among different countries, national groups and income classes. However, occupational work for women has become an essential element of the modern economy, and it is not easily dispensable.

2.7.4 Occupational Statistics

The statistical work on occupations deals with a descriptive analysis of the economic and social composition of population. Statisticians have simplified the concept of occupation on the basis of increasing dominance of the income seeking element in occupational affiliation and the changes in methods and norms of taxation. Modern occupational statistics list the income yielding occupations of the individuals composing the populations, and group them by geographical divisions and industrial income sources like agriculture, trade, mining, manufacturing and public employment. These statistics, therefore, give a picture of the type of economy (such as agrarian, industrial, rentier, etc.) prevailing in a country and give some ideas of the comparative numerical strength of the labour and property income groups. The methodology of occupational statistics, however, has undergone continual changes from the time of the first official occupational censuses undertaken in large European countries at the beginning of the nineteenth century and in the USA, from the mid-nineteenth century. Such changes in the methodology reflect changes in social theory and social ideals as well as changes in occupational realities.

Inspite of the fundamental similarity in the statistics of modern industrial countries, there are important differences in the method of data collection and analysis, which affect the results of occupational census. One important difference lies in the listing of *secondary* occupations in addition to the chief occupation. This method offers several advantages since the fully developed capitalism of the twentieth century seeks to utilise every opportunity for gain and often results in a complex and changing occupational life for the worker. The inclusion of secondary occupations in the census shows how large a portion of the population pursues several occupations. Another advantage of including secondary occupations is that it gives some indications of the degree of participation in agricultural work by women and older children whose main occupation in general is the non-remunerative household work. Another important difference in the collection of occupational data lies in the very basis of inclusion or exclusion of various groups. Virtually, all countries exclude such activities as the holding of honorary position, amateur pursuit, professional crime, etc. In some countries, economic characteristics of an occupation are given more importance. In the USA, only those who are gainfully employed are questioned as to their gainful occupation. In Germany, where the occupational census is connected with the general population census, it covers the total population without the limitation of age or occupational activity. In the countries where the occupational census is taken separately, a minimum age limit is taken into account for the purpose of occupational activity. This may be ten years as in the case of the USA, the UK and Italy; fourteen years as in India or fifteen years as in Norway.

The strictness of the definition of occupation for the purpose of census is more important for the ratio of those with an occupation to those having none. The residual group under "no occupation" includes such diverse groups as rentiers, pensioners, unemployables of various types (e.g., invalids, insane and prisoners) as well as students and inmates of social welfare institutions. In connection with the relation between the two major groups of those with an occupation and those without it, a quantitatively important question arises with reference to housewives in the city as well as married women and adult daughters on farms. The problem here is not merely one of the number employed in the "housekeeping occupation". It also involves the decision as to the value to be assigned to this occupation as compared with the various pursuits of women outside the home. Even, more difficult is the decision about the unpaid housework performed by adult daughters.

MODEL QUESTIONS

(Essay/Long Type)

2.1 Define and explain society.

2.2 Differentiate social organisation from social process.

2.3 What is social stratification ? Explain.

2.4 Explain the functionalist theory and the conflict theory with respect to social stratification.

2.5 Define and explain caste.

2.6 Explain the caste system in India and its origin.

2.7 "The caste system in India has undergone profound modification"-justify the statement.

2.8 Explain the concept of class.

2.9 Write in brief the historical evolution of class.

2.10 Explain the factors behind the formation of classes.

2.11 Define and explain culture.

2.12 Differentiate in brief culture from tradition, knowledge, magic and mythology.

2.13 How is the culture related to religion and arts ? Explain in brief.

2.14 Explain the concept of occupation.

2.15 Write in brief the historical evolution of occupation.

2.16 Give an account of the current trends with respect to occupation.

2.17 Comment on occupational statistics.

(Objective/Short Type)

2.18 What is the main difference between a caste and a clan ?

2.19 What are the two major, basic physical features of Aryans ?

2.20 Who are burghers ?

2.21 What is the main function of art ?

2.22 Name two major groups of people who have no specific occupation.

2.23 The Institution of Engineers(India) can be called :

    (a) Society   (b) Association

    (c) Trust   (d) None of these.

2.24 Solidary and antagonistic are two types of :

    (a) Tradition   (b) Culture

    (C) Social process   (d) Social organization.

2.25 The caste gives almost the best sense of solidarity to a :

    (a) Parsee   (b) Hindu

    (c) Christian   (d) Buddhist.

2.26 Match the following :

    (A) Untouchability   (a) Division of labour

    (B) Culture   (b) Dr. B.R. Ambedkar

    (C) Occupation   (c) Magic

    (D) Mythology   (d) Civilization

2.27 Match the following :

(A)	Classless society	(a)	Exogamous
(B)	Black magic	(b)	Art
(C)	Clan	(c)	London
(D)	Magic	(d)	Socialists

2.28 Write true or false :

Power refers to the style of life and the respect one commands in society.

2.29 Name the odd ones :

(a)	Friday	(b)	Saturday
(c)	Number 13	(d)	Broken mirror.

2.30 Name the odd ones :

(a)	Wealth	(b)	Cultural heritage
(c)	Classless society	(d)	Prestige.

2.31 Fill up the blanks :

The four main castes in India are

(a)	Brahman	(b)	Shudra
(c)	_____ and	(d)	_____ .

Societal Dynamics

3.1 MOBILITY

3.1.1 Types of Mobility

The term *"mobility"* or *"social mobility"* in its widest sense refers to any movement of individuals, families or social groups among different sectors of society. The movement in the same country from one occupation to another or from one religion to another is a kind of mobility at the national level. On the international scale, migration of people from one country to another is a very important type of mobility. In any discussion of mobility, a distinction is usually made between the so called *"horizontal"* and *"vertical"* mobility. The horizontal mobility involves no change in the position of the individual or the group in the social hierarchy. When a worker moves from one factory to another, or when an engineer takes a position in another company, there is no significant change in his social status. This is an example of horizontal mobility. The vertical mobility, on the other hand, involves a change of social level in the process of movement. If a worker becomes a wealthy businessman, his position in the class system changes quite radically. This is an example of vertical mobility, (upward). It may be noted here that vertical mobility may involve either *upward* or *downward* movement. An example of downward mobility is a member of an upper class who is dispossessed of his wealth in a revolution and is forced to enter into a manual occupation.

   Modern sociologists have concentrated their attention mainly on the *upward* mobility. This is due to the fact that they have been preoccupied with the question of equality of opportunity. In recent years, it has been suggested by many social theorists that the degree of *downward* mobility might be a far better indicator of the "openness" of a particular society. The reason is that downward mobility can clearly show the extent to which it is possible for the privileged classes and groups to maintain their status and pass on to their descendants the advantages that they enjoy. Though sociologists have been mostly preoccupied with the mobility of individuals, it is easy to see that whole families, groups, and even classes may, at certain times, change their position in the social structure. One of the earliest writers to bring out the diverse aspects of vertical social mobility was the economist Joseph Schumpeter, who in 1927 analysed and illustrated what he termed the "*rise and fall*" of individuals, families, and whole classes within the class structure. In studying the movement of individuals, sociologists distinguish between the following :

(1) Intragenerational Mobility.
(2) Intergenerational Mobility.

The Mobility is said to be *intragenerational* when an individual moves vertically upward or downward within his own adult lifetime. On the other hand, if vertical movement represented by a change in the social level occurs from the parental to the filial generation (mostly from father to son), it is known as the intergenerational mobility. In recent years, comprehensive national studies of mobility have been conducted in many countries, but such studies have dealt almost entirely with intergenerational mobility, investigating changes in occupation between father and son. Another limitation of these studies is that they have concentrated heavily on the educational opportunity as a major factor influencing upward mobility. In contrast to the mobility of individuals, the rise and fall of families, groups or classes is relatively more difficult to study, but such movements can be documented in other ways. An important example of the upward mobility of families is the emergence of new ruling dynasties. Other examples can be found in the rise and fall of family business or politically influence. Such mobility of particular social groups usually occurs as a result of economic, political and cultural influences.

3.1.2 Historical Studies of Mobility

The problem with historical studies of mobility is that they largely provide illustrations of various types of mobilities but do not show the extent of it due to lack of data. Such historical studies do not give the actual number of individuals or families who change their position by upward or downward movement in the social hierarchy over a given period of time. As a result, historical comparisons of mobility are quite difficult and largely speculative. In fact, there are many difficulties even in examining the trends of mobility in more recent industrial societies. Nevertheless, most sociologists agree that the vertical social mobility (especially that of individuals) is greater in modern industrial societies than it was in earlier societies. On the other hand, modern social research shows that, even in industrial societies, mobility is limited and that there are no substantial differences between different industrial societies in the extent of mobility in spite of the general belief that social mobility is greater in the United States than in the European countries. The movement of individuals from the working class into the upper class is generally *rare* in *all* societies. Comparisons among industrial societies show that the greater part of the vertical social mobility is a short-range mobility.

Modern investigations suggest that one should be cautious in the interpretation of historical accounts of social mobility and undue prominence should not be given to exceptional cases of upward or downward movement. It is quite likely that, in all societies and at most times, there is considerable amount of stability. When this stability is not maintained by any formal or legal sanctions, it is ensured by the inheritance of property, educational advantages or political influences.

3.1.3 The Process of Mobility

The nature and degree of vertical mobility in a society are governed by a number of factors. One *universal* factor is the occurrence of individuals with exceptional endowments such as intelligence, physical strength, beauty, business acumen, etc. In the past, beautiful women rose to social eminence as the mistresses of kings and nobles, and in more recent times, as film stars. Similarly, men (and to some extent, women too) have risen in the social hierarchy by accumulation of wealth, attainment of political or military power, and intellectual or artistic achievements. Manifestations of such personal qualities in exceptional amounts are quite limited and, so are the social advantages that they bring.

Vertical social mobility resulting from exceptional personal qualities is governed by many social factors. In the first place, the open or closed nature of the class system in a particular society has a powerful influence on the vertical movement of individuals. An individual in a *close* system encounters many obstacles if he seeks to escape from his social position as a slave, serf, or a member of a lower caste. The ideology that upholds such a system usually tends to inhibit the development of talent and ambition at the lower levels of the society. In the more *open* class systems of modern societies, there are no formal restrictions on upward vertical mobility. However, talented individuals from a lower social class have to overcome many difficulties, arising mainly from poverty and lack of education, in order to succeed and move to a higher social class. In contrast, less talented individuals from the upper classes are able to maintain their positions due to their inherited social advantages. The extent to which an individual talent will lead to upward mobility is also limited by the general orientation of the activities of a particular society. A primitive tribal society that lives by hunting or is engaged in an on-going warfare with other neighboring tribes will naturally place a high value on physical strength. Similarly, a nation engaged in imperial expansion and colonization will highly rate military qualities; and the nation that is mainly concerned with industrial and economic developments will attach the greatest importance to business skills.

The general changes in social structure has a profound influence on social mobility. For example, a revolution that dispossesses an existing upper class or a national liberation movement that overthrows foreign rule creates new opportunities for individuals, groups and even whole classes to move upward and occupy dominant positions. The mobility is also affected by more gradual changes in a society, especially if the changes occur in its occupational structure. Thus, the expansion of technical and professional employment involves a continuing movement out of manual work, and this accounts for a major cause for the upward mobility in modern industrial societies.

Mobility is high during a revolution and after a revolution, a system of social stratification may emerge and the rate of social mobility may decline. Similarly, during recession or depression, when the rate of growth slows down, opportunities for mobility are greatly reduced. In such circumstances, the possibility of upward mobility will largely depend upon the extent of downward mobility. Another factor having an influence on the mobility, when there is little economic development, is that of *differential fertility*. If the upper and middle classes limit their families, the vacant places may be filled by individuals who rise from the lower classes. Such differential fertility has rarely been a major influence on social mobility. Upward social mobility has often been promoted by international movement of population. A very important example of this phenomenon is the colonial expansion of many European countries from the sixteenth century onwards, which provided opportunities for individuals to move upward in society by enriching themselves as traders or settlers and, at the same time as they subjugated other people in Asia, Africa and America. At a later stage, with the creation of new societies by European settlers in their adopted countries (especially in North America and Australia), fresh opportunities for mobility were provided by large scale immigration.

3.1.4 Consequences of Mobility

The vertical movement of a large number of people up and down the social hierarchy tends to break down the exclusiveness of the social class and creates a more uniform national culture. Hopefully, this may also lead to reduction of class conflict, or at least class prejudices. Many sociologists have claimed that the lesser degree of class consciousness in the US society, as compared to the societies of European countries, is due to the highest rate of social mobility in the US. It may be noted that the widespread *belief* that opportunities for upward mobility are greater in the US than in other societies may itself have had an important influence on the lesser degree of class-consciousness in the US. The other side of the coin of social mobility is the argument that preoccupation with vertical movement *reinforces* the class system. Thus, the individuals who are concerned to rise or atleast to avoid falling in the social hierarchy accept and even emphasize the importance of class and status distinctions. Another consequence of mobility is the more effective use of individual talent and ability, and this is generally considered beneficial from the point of view of the society as a whole. The reason is that if the individuals are confined to the social segment in which they are born, many useful talents will remain undiscovered and unused. In modern industrial societies, the expansion of education has been stimulated by the desire of the governments to provide opportunities for the development of all the abilities in the population, although most of the countries are far from achieving this goal.

Vertical mobility may also have some undesirable consequences. Vertical social mobility, both upward and downward, produces strain in the individuals striving for success and adapting to new social milieus. Such mobility may also be disruptive of families and local communities, (e.g., in Kerala, where thousands of talented individuals have gone to work in the Middle East countries, disrupting their family life). A high rate of vertical mobility may produce in a society the condition that the French Sociologist Durkheim called *"anomie"* (meaning normlessness) and the resultant disorientation and anxiety. Under this condition, there is insufficient regulation of the individual behaviour, and the individual suffers from the "malady of infinite aspiration". The existence of such stress and strain may also give rise to a higher incidence of mental illness among the highly ambitious and highly mobile individuals.

3.1.5 Channels of Mobility

Many social institutions like army, church, school, political party and occupational organisation serve as the channels of vertical mobility through which individuals ascend or descend the social ladder. In any society, at a given period of time, one of these institutions may play a dominant role, (the army in time of war). With the exception of the period of anarchy, vertical mobility is strongly controlled by the elaborate social machinery of testing, selection and placement of individuals with regard to various social positions. The family, the church and the school test the general intelligence and character of the individual according to their standards. When the individual enters an occupational institution, he is tested for the specific ability and skills necessary for the successful performance of definite functions. These institutions, therefore, serve as social sieves. They perform not only the educational and training functions, but the selective functions as well.

The process of vertical mobility exerts a number of important influences upon social life. Intensive vertical mobility increases plasticity and versatility of behaviour and stimulates progress in thought, discovery and invention. On the other hand, too high a rate of social mobility seems to increase mental diseases when the individuals face difficulties in adaptation to the new situation. Vertical mobility makes the social structure elastic, breaks the isolation created by class and caste and stimulates rationalism. Its direct and indirect influences on all aspects of social organisations are very potent, but highly complex.

3.1.6 Current Trends in Mobility

The economic development of the Western industrial nations after the World War II and the provision of more elaborate welfare services have generated significant changes in the class system. These changes have been interpreted in many different ways by various sociologists. According to some authors, there has been a general diminution in class differences. This has resulted in a higher level of living, greater social mobility and a limited redistribution of wealth and income. These social changes are reflected in a decline in class conflict. This interpretation leads to the conclusion that the Western societies are moving in the direction of relative classlessness or that they are becoming predominantly middleclass societies. In sharp contrast to this interpretation, other social scientists have argued that the social changes since World War II are leading to the formation of new social classes like a new upper class consisting of managers and organizers of production in both public and private sectors.

3.2 INCOME DISTRIBUTION

3.2.1 Concept of Income

Income results from the services rendered to the society by an individual or an organisation. Service may be defined as the creation of desirable events/ results or the avoidance of undesirable events/results. The value of an income is the value of the services. For all practical purposes, the value of an income is measured in terms of money. Under services, one may include the following :

(1) The benefits from a property right such as the interest yield from a bond or a fixed deposit in a bank.

(2) The benefits derived from objective instruments such as the shelter offered by a dwelling.

(3) From the co-operation of individuals with such objective instruments (the transport service of a railway).

(4) The services rendered by individuals whether they are manual labourers or highly qualified professionals.

The total income of an individual or personal income is the total money value of the services received by him from all sources during a given period of time, usually in one year. Similarly, a person's net money income is the sum of all his money receipts less the money invested by him. The income of society as a whole is the total money value of all the services received by the members of the society from all sources. In case of an individual, his "real income" is the money value of his expenditures on food, clothing, shelter, amusements and other miscellaneous services.

One important result of these principles of income calculation is the exclusion of capital gains from the income. The capitalisation, at any point of time of the income expected in the future, is not itself an income. By the same token, an increase in the capitalisation from one point of time to another is not income except potentially. Thus, if a fixed deposit of Rs.1000.00 in a savings bank is earning Rs.100.00 a year as interest and if the depositor withdraws his interest earnings every year, the actual income from his fixed deposit is Rs.100.00 a year and the principal sum of Rs.1000.00 is the capitalisation of this income. On the other hand, if he does not withdraw the interest, this amount is merely accrued and becomes capital gain rather than an income.

Certain peculiarities attach to income from labour as contrasted with income from property. Thus in computing the income of a domestic help, it would be necessary to include not only his wages but also his perquisites like boarding and lodging since these are a part of his compensations, in addition to money wages. Moreover, accountants do not usually depreciate the income of a working man as they depreciate in case of a machine, for they cannot easily appraise the value of a human being. More complications arise due to the fact that income must be defined for the purpose of taxation. The kind of income that is taxed is largely determined by the economic organisation of a country and the fiscal needs of its government.

3.2.2 National Income

National income may be defined provisionally as :

(1) The net total of commodities and services produced by the people comprising a nation.

(2) The total of such commodities and services received by the nation's individual members in return for their assistance in producing commodities and services.

(3) The total of goods and services consumed by these individuals out of the receipts thus earned or

(4) The net total of desirable events enjoyed by these individuals in their dual capacity as producers and consumers.

Defined in any of the above mentioned manners, national income is the end product of a country's economic activity. As a result, national income reflects the combined play of economic forces and serves to appraise the prevailing economic organisation of a country. Figures for per capita income, especially when adjusted for differences in purchasing power of money, measure the economic welfare of a country. A continuous series of annual estimates of either the total or per capita national income suggest whether the nation tends in the course of time to grow richer or poorer and how rapidly this change takes place.

3.2.3 Distribution

3.2.3.1 *Problem of Income Distribution*

It is not enough for a country to attempt to increase its national income by development programmes. The national income must be increased; but it is also necessary to ensure that it is equitably distributed among various sections of the society. Inequality of income is an important feature of capitalist economies. Even socialist and communist countries, who have established systems for the purpose of reducing inequalities of personal income, have failed to attain this equality. In pre-historic times, there was no need for a policy on income distribution since man led a nomadic life, always in search of food. Income distribution came into force during the feudal system and attained great importance with the advent of the Industrial Revolution. Government policies to ensure a fair distribution of personal income are among the most controversial and difficult issues of public policy.

3.2.3.2 *Causes of Inequality*

The major components of personal income are labour earnings (salary or wages), property earnings (rents, interests and dividends) and government transfer payments. Disposable personal income consists of the personal income minus any taxes paid on it. Wealth (or "net worth") consists of the total value of financial and tangible assets minus the amount of money owed to bankers or other creditors.

The major causes of inequality of incomes in an economy are:

(1) Inheritance: Some persons are born with a silver spoon. Rich inheritance gives them a good start in life. Some persons are born landless; others inherit thousands of acres. Some parents die under debt leaving the burden

of debt on their children while others leave huge cash and properties to their heirs. So long as the system of inheritance continues, inequalities are bound to continue.

(2) System of Private Property: Under the system of private property, a person is free to earn, free to save and free to own property. Once a person earns and acquires property, his property starts earning for him (by way of rent, interest, etc.). That is why some earn more and others earn less and differences in property lead to difference in income. Property is one of the major causes of the inequality of income.

(3) Differences in Natural Qualities: No two persons have the same natural qualities. Some are more gifted than others. Persons who are endowed by nature with superior intelligence, better physique and greater capacity for hard work can easily surpass others in the race of life. Some inherit a feeble mind in a feeble body and they are left behind.

(4) Different in Acquired Talents: To some extent, environment makes the man since natural (or inborn) qualities are considerably modified by environment. A child may be highly intelligent, but if he or she is not lucky enough to receive proper education or training, the latent abilities remain mostly undeveloped. On the other hand, a child of even mediocre nature abilities can do better if he or she is properly brought up and educated. Professional education, for example, improves a person's earning capacity.

(5) Lack of Opportunities: Some persons are lucky enough to get a good chance, and they may make the most of it. It is well known that underdeveloped regions (like Uttaranchal, Jharkhand, Chhattisgarh and Nagaland) do not offer good opportunities for employment, where as developed regions (like Punjab, Maharashtra and Gujarat) have ample opportunities.

3.2.3.3 Consequences of Inequality

Inequality of income leads to serious economic and social consequences. Some of the major consequences of uneven income distribution are as follows:

(1) Class Conflict: Inequitable distribution of income and wealth has divided the society into two classes, the "haves" and the "have-nots", which are forever on the war path. This class conflict leads to social and political discontent.

(2) Political Domination: The rich dominates the political machinery and uses it to promote his own interests. This results in corruption and social injustice.

(3) Exploitation of the Poor: The rich exploits the poor economically, socially and politically. The awareness of this exploitation may lead to

political awakening, agitation and even political revolution. Inequality of income is an important cause for social and political instability.

(4) Creation of Monopolies: Unequal distribution of income promotes monopolies. The monopolies can crush small enterprises and change unfair prices.

(5) Suppression of Talent: It is not easy for a poor person to make his way in life, no matter how talented he or she may be. It is a great social loss that highly brilliant but poor people are not able to make their full contribution to the society and the nation.

(6) No Real Democracy: Democracy is a farce when there is a wide gulf between the rich and the poor. There can be no real democracy and political equality without economic equality.

(7) Moral Degradation: Unequal distribution of income leads to moral degradation of the society as the rich are corrupted by vice and the poor are demoralized by lack of economic resources. The economic inequality corrupts the rich and degrades the poor. It becomes almost impossible for the poor to retain their honesty and integrity when they see the corrupt and rich people rising in life.

3.2.3.4 *Measures to Reduce Inequalities*

In the present era of social and political awakening, it has become a major plank of political policy to reduce the inequality of income distribution, if not eliminate it. After the independence, India decided to set up a socialistic pattern of society. With this end in view, the Government of India strives to prevent the concentration of wealth and income in a few hands. Some of the measures to reduce inequality in the distribution of income and wealth are:

(1) Fixing Minimum Wage: The first step in the direction of a more egalitarian society is to guarantee each citizen a minimum wage, consistent with a minimum standard of living. The Minimum Wage Act was passed in India in 1948. In pursuance of this Act, minimum wages are fixed from time to time for agricultural labour and other workers.

(2) Social Security: An important measure for a more equitable distribution of income is the introduction of a comprehensive social security scheme assuring each citizen a minimum standard of economic welfare. Such a social security scheme must include provision for free education up to cer-tain level, free medical and maternity aid, old-age pension, unemployment benefits, compensation for sickness and accidents, provident fund and group insurance schemes. In this way, substantial benefits can be assured even to persons whose incomes are low. Social services like public parks, libraries, museums, community halls and community TV sets may be pro-

vided on a liberal scale so that poor are able to enjoy many of the amenities available to the rich.

(3) Equality of Opportunity: The government may devise suitable means to provide equal opportunities to both the rich and the poor in getting employment or getting a start in trade or industry. For example, the government may institute a system of liberal scholarship, stipends and low-interest loan so that the poor can acquire higher education and technical skills. In India, many concessions are offered to the scheduled castes, scheduled tribes, other backward classes and persons living in backward areas to reduce inequality in the society.

(4) Steeply-Graded Income Tax: All possible fiscal devices may be adopted to bridge the gap between the rich and the poor. One such device is the steeply-graded (i.e., progressively higher) income taxes. This can prevent to some extent the rich getting richer. Other direct taxes like the super tax, the excess profits tax, the capital gains tax and limits on dividends may also be used for this purpose.

(5) Steep Estate Duty: In order to prevent the perpetuation of inequality from generation to generation, steeply-graded estate duty, death duty and succession taxes may be imposed. In 1964-65, and again in 1966-67, the rates of estate duty were made steeper in India, rising up to 40%.

(6) Ceiling on Property: With a view to reducing inequalities between the big and small farmers, ceilings on agricultural holdings may be imposed, as has already been done in India. The main purpose of land ceilings is to bring about a wider ownership and use of land. Similarly, a ceiling on urban property may be imposed so that the inequalities in urban areas can be reduced.

3.2.4 Statistical Analysis of Income Distribution

In the first place, there has been an attempt to establish a functional relationship between the size of income and the number of recipients. Secondly, an attempt bas been made to summarise the income distribution by a single measure of the inequalities of income. The most famous attempt in the first direction is the **Pareto's law**. The law, in its most dogmatic form, states that the distribution of income in the upper ranges of income tax payers shows a linear relationship. Mathematically, the Pareto's law may be stated as:

$$\log N = \log A - \alpha \log \chi \qquad\qquad 3.2.4.1$$

where χ is the income size, N is the number of individuals having an income equal to or larger than χ and A and α are constants (found from empirical statistics by fitting the data to the straight line given by the above equation).

It has been found that the constant α (the slope of the straight line) is approximately equal to 1.5 in all countries at present. In addition, *all* ranges of income distribution follow the same linear relationship for all countries at present. It follows, therefore, that because of the unchanging and unchangeable nature of the whole range of income frequency distribution, economic welfare can be increased only through an increase in the total amount of income. It is obvious that the Pareto's law is of great importance for major questions of economic theory as well as economic policy. Many economists and statisticians have directed their attention towards testing of its validity. Results of such cumulative analysis have shown that the Pareto's law is quite inadequate as a mathematical generalisation. Because of the heterogeneity of the frequency distribution curve (due to grouping together of income from various economic categories), it is unlikely that any mathematical law describing adequately the entire distribution of income can ever be formulated.

Other attempts to substitute for the *linear distribution* with another mathematical expression have also been found unsatisfactory for describing the distribution of income. However, a French economist, R. Gibrat, has obtained successful descriptions of a large number of frequency distributions of income by using a modification of the normal distribution curve of errors. The curve employed by Gibrat is:

$$y = \frac{1}{\sqrt{\pi}}\, e^{-z^2} \qquad\qquad 3.2.4.2$$

and
$$Z = a \log (\chi - \chi_0) + b \qquad\qquad 3.2.4.3$$

where y is the number of income recipients, χ is the variable size of income, $(\chi - \chi_0)$ is a selected income constant, while a and b are constants to be found from empirical statistics. The assumption in which eq. 3.2.4.2 differs from the normal distribution curve is that the effect of each of the numerous contributory factors is not independent but proportional to the effect of other factors.

3.2.5 Measures of Inequality of Income

The more fruitful developments in the direction of summarising inequality of income by a single measure has yielded numerous measures. Such single measures of inequality of income broadly fall into four groups :

(1) The measures derived from a specific type of mathematical equation and hence, contingent upon the goodness of fit of the curve implied by that equation.

(2) The measures of the mean deviation type, available in the statistical theory of frequency distribution and applicable to diverse types of distributions.

(3) The measures of mean difference types.

(4) The measures constructed by using definite theoretical criteria in regard to welfare equivalents of individual income.

In the first group, there are three important measures of inequality. The coefficient α of Pareto's equation has been employed as a measure of inequality. The steeper the slope (the larger the numerical value of α), the smaller the inequality of income.

A second measure of inequality is Gini's index of concentration δ, which is derived from another equation of income distribution :

$$\log N = \delta \log S - \log K \qquad\qquad 3.2.5.1$$

N is the number of individuals whose income is *above* a certain size, S is the sum of incomes (each greater than the certain size), and δ and K are constants to be determined from empirical data. It may be noted that N is a function of the *sum* of incomes greater than a certain size rather than a function of that income size itself, as in the case of the Pareto's law. The relationship between Pareto's measure α and Gini's measure δ can be expressed by the equation

$$\delta = \frac{\alpha}{\alpha - 1} \qquad\qquad 3.2.5.2$$

A third measure of inequality of income may be derived from the curve employed by Gibrat . This measure is taken to be equal to $100/\alpha$.

Of the dispersion measures developed in the statistical theory of frequency distribution, the average and the standard deviation naturally suggest themselves as indices of the inequality of income. The resulting relative measures of dispersion can be obtained from a frequency distribution in which the class intervals of income size are taken in absolute figures or in logarithms. The advantage of the latter procedure arises from the fact that the positive skewness characterizing frequency distributions of income is reduced by taking the income variable in terms of logarithms.

The mean difference of incomes is given by the arithmetic average of differences(taken without regard to their positive or negative signs) between all possible pairs of incomes. This measure was suggested by Gini and it is known as the "*ratio of concentration*". Another widely known measure of inequality, which is related to Gini's ratio of concentration, is known as "*Lorenz curve*". In the Lorenz curve, the cumulative percentages of total income (Ic) are plotted along the X–axis, while the cumulative percentages of population (Pc), from the poorest to the richest, are plotted along the Y–axis. In the Lorenz curve, an equal distribution of income (total absence of inequality) is represented by a straight line passing through the origin and having a slope equal to unity, as shown by the straight line A in Fig. 3.2.5.1.

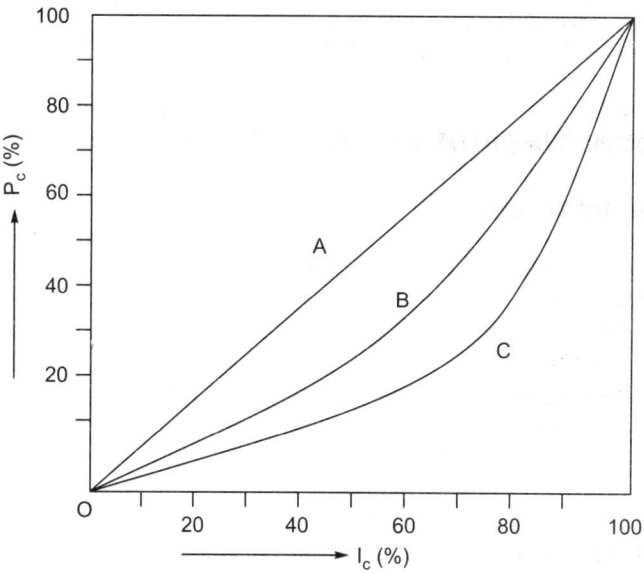

Fig.3.2.5.1 Typical Lorenz curves (Ic=cumulative percentages of income and Pc = cumulative percentages of population)

Empirical distributions of income, on the other hand, usually appear as concave hyperbola, as shown by the curves B and C. The existing inequality of income is measured by the area between these hyperbolas and the straight line. The curve B shows less inequality of income, while the curve C indicates greater inequality. The variety of methods devised to measure the inequality of income illustrate the profusion of various aspects. It also suggests a high probability of divergent results from the analysis of one and the same data. In addition, this lack of agreement as to the precise aspect of inequality to be studied is usually accompanied by paucity of adequate empirical data on income distribution. Considering these two factors, therefore, one would expect to find few definite conclusions as to the trends or differences in the inequality of income. On the contrary, one usually finds a profusion of contradictory generalisation, which are often obvious results of pressure to respond somehow to a problem vital to social policy. How divergent and unreliable such inferences are may be illustrated in the case of former Prussia, which is one of the few countries for which empirical data on personal distribution of income are available for a long period. From the income data for Prussia for the years 1875,1896,1913 and 1919, Prokopovich concluded that the inequality of income was increasing during the period 1875 – 1919. A contrary conclusion was derived from the *same* data by Helfferich, who claimed that no tendency towards a greater concentration of income was observable. On the other hand, Angelopoulos studied the data for the years 1896, 1914 and 1926 and inferred that the inequality of income had diminished during the period 1896 – 1926.

3.3 SOCIAL TENSIONS AND THEIR CAUSES

3.3.1 Social Tensions

Social tensions usually result from the conscious pursuit of exclusive values by individuals and/or social groups. The individual members and groups always widen or narrow their claims on society for life, liberty of movement, property and other desirable things by competing against one another. Competitors become self-conscious rivals, opponents or even enemies. The relative position of an individual in relation to the current value in a society is controlled by influences of which he is mostly unaware. On the other hand, everyone believes in some measures that his personality and life can and should be protected from the encroachment by others and that it can be enhanced at the expense of others. Thinking along these lines by individuals and groups leads to *social conflict*, which may involve the defence of what already has or the acquisition of what one does not have. Acquisition may mean taking away of that belongs to another or the appropriation of that which another would like to have. The defensive, destructive and obstructive aspects of a social conflict become entangled with one another in every crisis and create social tensions.

Exclusive values may be pursued by individuals and groups by various means which may vary all the way from physical violence to persuasion and thus include the whole range of techniques used for social control. Conflict among human beings differs from the struggle for survival among animals in the diversity of the means employed by the former. Each animal is bound by its very structure to a few stereotyped means for coping with a conflict situation. On the other hand, the nature of man is so plastic that rich variations can be found within the human species among the means employed to deal with conflicts. Thus, the place of physical combat is usually taken by argument, admonition, ridicule, litigation, etc. and these are the functions of cultural setting in which conflicts occur.

Social conflict is a mode of registering or consummating social change. As a result, conflict is more acute in case of rapid social changes, when the vested interests of the old social order stand out against the pressures from the new. A social crisis may bring about a willingness to come to terms with the requirements of new realities. Thus, strikes, boycotts and wars frequently lead to a more permanent organisation of interests and social tensions and subside after prolonged agitations and protracted litigations. Social crisis may also result in the obliteration of one or both of the contending parties. It is not always true, however, that a particular social conflict can produce a relatively abiding settlement. It has often been said that vital issues are never decided; they are merely superseded. There are numerous causes of social tensions. The most important ones are:

(1) Class struggle.
(2) Political corruption.
(3) Crime.
(4) Competition.
(5) Unemployment.
(6) Heterogeneous population.
(7) Racialism.
(8) Religious intolerance.
(9) Linguistic intolerance.
(10) Uneven distribution of wealth.
(11) Rapid social change.
(12) War.

Among the above twelve causes of social tensions, the first five are very visible.

3.3.2 Class Struggle

Class struggle is a form of social conflict, which creates tension in the society. Credit for the theory of class struggle goes to Karl Marx and Friedrich Engels. They formulated the doctrine of class struggle in the *Communist Manifesto* in 1847. Marx's theory of class struggle asserts that, in the course of making a living and utilizing technical knowledge and industrial equipment, the members of society become segregated into classes which carry on different functions in industry and, as a result, occupy different positions in the social organisation. Among these classes, there arises an antagonism of interests, which may lead to a struggle. The modern capitalistic regime does not abolish the class struggle; it merely creates new classes and intensifies the struggle among the classes. With the development of capitalism, society splits up more and more into two hostile camps - the bourgeoisie and the proletariat. The bourgeoisie attempts to convert surplus value into profits, while the proletariat tries to resist it. In the ensuing struggle, the workers realize that the power of the bourgeoisie rests on the ownership of the means of production and that economic exploitation can be ended only through establishment of a socialist society based on collective ownership. On the basis of his theory of class struggle, Marx predicted the inevitable victory of the proletariat with assumption of certain inherent tendencies in capitalism, viz., the rapid concentration of industry, the disappearance of the middle class and the growth and increasing misery of the working class.

In any society, the process of production is simultaneously a process of economic exploitation. Those who do physical work receive less than they produce, not only because a portion of it is necessary for the extension of production, but also because they have to support the owners of the means of production. The resulting antagonism finds its expression in a struggle for distribution of the total national product. As this struggle becomes conscious, it gives rise to class conflict and creates social tension. The dominant minority tries to maintain and extend the opportunities for exploitation, while the exploited majority continually strives to liberate itself. Gradually, class interests intermingle with political, religious and even scientific and technological interests. Since the power of the ruling class is always concentrated in the organisation of the state, the oppressed class aims directly against the state. Every class struggle is a political struggle, which aims at the abolition of the existing social order and establishment of a new social system. In order to prove capable of establishing a new social organisation, the class must possess certain essential characteristics:

(1) It must be economically exploited and politically oppressed.
(2) It must be a producing class.
(3) It must be welded together by the conditions of its existence.
(4) It must form a large mass or a majority of the population.

According to the theory of class struggle, all these characteristics are found in the industrial working population. It is for this reason that the industrial proletariat represents the only class which can carry out a complete social revolution and can reorganise society on the basis of the socialist ideal. The peasantry lacks most of the above traits. On the other hand, the peasantry forms a large portion of the population in most countries. So, it is essential that the industrial workers should ally themselves with the poorer elements of the peasantry to carry out the social reorganisation.

The classical theory of class struggle, as outlined above, has encountered many difficulties. No definitive division of society can be made into classes on the basis of the so-called factors of production, the law of the division of labour or the sources of income. In modern society, the struggle by economic and social groups is fragmentary and intermittent. It is generally concentrated around major issues of immediate importance. It is also inevitable that economic interests should be overshadowed from time to time by cultural, religious or racial factors. Nationalism, both as economic and cultural phenomena, tends to offset the formation of classes. Economically, each nation is likely to regard itself as one unit with common interests as against other nations. Culturally, each nation tries to consolidate its economic coherence by means of national ideals which permeate all social groups. To the extent to which it manifests itself, the struggle of economic groups is a potent factor of social change. Manifestations of group conflicts like strikes reveal stagnant conditions and serve as a stimulus to their elimination or amelioration. On the other hand, when such conflicts in the industrial world are concerned merely with group shares in the distribution of wealth, they may result in social detriment. In large social transformations, the struggle of classes may lead to social disintegration. This is true especially when the class struggle is carried out among social groups which are in early stages of economic and intellectual development. Such class struggles have often resulted in social crises.

3.3.3 Political Corruption

Political corruption is generally the misuse of public power for private profit. Although no political group has been entirely free from jobbery, corruption has not prevailed to an equal degree at all times or under all conditions. It has responded to various opportunities for the misuse of power created by various governmental institutions. Any analysis of the prevalence of political corruption, therefore, must regard it as a phenomenon of group psychology, conditioned by the entire cultural setting of the group. Thus, no remedial programme to eliminate political corruption can be fully successful if it limits itself to the prosecution of individual offenders or even to administrative organisations. Political corruption requires the existence of public officials with power to choose between two or more courses of action and the possession by the government of some power or wealth which can be used for private gains. These two prerequisites are found even in relatively simple primitive societies. However, corruption was generally scarce in primitive societies, primarily because of the dominance of tradition in determining conduct and the promptness of punishment.

It was the British historian Lecky who remarked in his book *History of European Morals* that "It is possible that the moral standard of most men is much lower in political judgments than in private matters in which their own interests are concerned". It is doubtful, however, that political morality is worse than average business morality. In modern societies, business and government are increasingly placed in direct relationship with each other through various government contracts, which are becoming more and more important with the growing scale of governmental activity. Especially conducive to corruption are contracts for military supplies, particularly in times of war. In modern times, war involves huge expenditure under conditions of relaxed public control since public attention is focused on supplying the soldiers with what ever they need regardless of cost. After the war, surplus war materials are disposed of at a fraction of their cost. Thus, recognition by businessmen of the opportunities for tremendous profits during and after the war increases the strain on the honesty of public officials.

It goes without saying that corruption is not restricted to any particular country or society. Corruption is an international phenomenon. However, developing countries are generally found to be more corrupt than developed ones. A recent survey of 50 important nations of the world has shown that India, Pakistan, Nigeria, China and Russia are among the most corrupt countries in the world, while Switzerland, Singapore, New Zealand and the US are among the least corrupt countries.

The exposure of corruption may come through individuals (investigative journalists) or organisations. Under the party system of governments, exposures of political corruption result from the desire of opposition parties out of power to obtain power. In many Latin American countries, where elections may not be very fair, revolution is the only way to accomplish the change of government. While some improvements in conditions often result from such revolutions, corruption generally continues for the benefits of new governments. No particular system of government is a guarantee against political corruption. Corruption can be eliminated only where the power to do so is linked with the desire to crush corruption. But power itself is an inducement to corruption. As the saying goes, "power corrupts and absolute power corrupts absolutely". Little progress in getting rid of corruption can be expected where those in power either benefit from corruption or fail to recognise the necessity for its elimination. Ignorance of the existence of corruption, widespread benefits derived from the lavish distribution of goods, political indifference (or lack of political will) and materialistic culture are the major factors which contribute to corruption in the political life of a nation and create tension in society.

3.3.4 Crime

Antisocial conduct (known popularly as *"crime"*) may be regarded as a universal phenomenon or a function of group-life. Its extent, particular form it takes and the reaction it provokes are intimately dependent on the cultural status and social organisation of the group. The term "crime" is used indiscriminately by the layman to denote antisocial, immoral or sinful behaviour. What the law calls crime is merely conduct which is declared to be socially harmful by the groups which are powerful enough to influence legislation in a state. The existence of crime in a community is a challenge to its members since crime is harmful for ordered social growth. Combating crime involves huge economic burden and enormous waste of misdirected energy. The progress of the science of human behaviour in the last several decades has increasingly centred its attention on criminal behaviour. As a result, the last half century has seen the development of the science of criminology. Official and private agencies for the study of criminals and their treatment have also multiplied rapidly during this period. The reason is that the task of dealing with crime and criminals is such that it demands a knowledge of facts upon which an intelligent programme of study and treatment may be based. Given the serious nature of crime and its adverse effects on society, much thought has been devoted to its control. Two general lines of thought have emerged for the purpose of controlling crime, the ameliorative and the repressive philosophies.

The *ameliorative philosophy* asserts that as the ranks of the middle class grow in a society, fewer and fewer people commit crime. Moreover, as the middle class grows, more people may become intolerant of crime and they commit offences more frequently. As a result, criminal acts that do occur can be treated more effectively. In order to control crime, it is necessary to bring about the middle class by upgrading socio-economically the lower class. In contrast to this, the *repressive philosophy* regards the modern society (especially in developed nations) inherently prone to criminal behaviour due to the decline of informal community control, greater secularisation and more egoistic moral code. Consequently, modern society may at best be expected to produce an affluent population lacking in personal restraint and civic consciousness. In the first place, punitive methods must be employed to redress offences and efficient organisation must be used to prevent crime.

Regarding enforcement of law, administration of justice and rehabilitation of criminals, the above two contrary lines of thought have given rise to three strategies (or policies) to combat crime. These three strategies are referred to as punitive, therapeutic and preventive respectively. The *punitive approach* regards the offender not only as justifying punishment morally, but as being also susceptible to deterrent measures. Punishment is, therefore, exercised to incapacitate the criminal for the present and to keep him away from engaging in any criminal act in the future. In many developed countries, the punitive method has lost ground to the other two ameliorative methods. As a result, the capital punishment has already been abolished in most countries. The *therapeutic approach*, on the other hand, deems the offender to be malfunctioning and is in need of psychological or sociological treatment. In the US, about 20% of the staff employed for law enforcement and administration of justice deal with corrections, with rehabilitation as their primary goal. The *preventive approach* centres on modifying the interaction of social and individual conditions to forestall offensive behaviour. This method calls for improvement of sociological environment, strengthening of social structure and development of conforming personalities. Education and recreation can play a vital part in this respect, and so can a wide range of other practices like vocational placement and community organisation.

3.3.5 Competition

Competition is an important term in social theory. It is by competition (of persons, firms, industries, nations, races or cultures) that the fittest survives. Competition arises out of rivalry, which is a universal fact in life and society. Rivalry is manifested in a struggle among germ cells among plants for sunlight and growth and among animals, for food and mates. Rivalry is evident in the striving in our daily life and appears in every social order under which men live. It is due to rivalry that conflicts occur among primitive tribes for hunting grounds and among capitalists to bag huge profits. In the march of history, a machine process wins its way against ancient crafts, a novelty called "business" displaces custom and authority in the control of industries and a modern creed replaces outworn dogma in domination over human mind.

The subject of competition has invited varied and often conflicting statements from various authors since a single explanation of a complex social phenomenon like competition is inadequate. Competition is hardly distinguished from free enterprise and capitalism. On the one hand, competition is the gigantic motor which causes the individuals to use their mental and physical powers to the best advantage to go ahead. Competition develops in the individual the habit of self-reliance. Competition has lifted the human race to a standard where the mode of living of common labourers in developed countries is more comfortable than the daily existence of ancient kings. On the other hand, competition is a nice new name for the brutal fact of all against all, without pity or mercy. Carried to its logical conclusion, competition may turn into anarchy.

In recent times, competition has invited attention of social thinkers and activists. One group demands that competition be preserved in all its native simplicity, while another group would like to eradicate the evil of competition and substitute it with a moral economic order. But since the shortcomings of competition lie in its specific performance, the general demand has been to mend the system and make it work. As a result, the arrangements of competition have been modified from within in many cases. Businessmen, for example, form trade, industrial or commercial associations (e.g., Truck Owners' Association) with an aim to make competition less ruthless and more rational. Similarly, farmers organise co-operatives (e.g., Anand Milk Union Limited, better known as "AMUL") to escape the tyranny of an uncontrolled market. Labourers form trade unions only to secure a bargaining power equal to that of their employers. Often, the state interferes in private matters for several reasons, to conserve natural resources, to ensure the quality of goods by standardisation, to fix hours of labour, to provide compensation for accidents and so on. But even the use of formal authority in such cases leaves the rival firms in an industry as free to compete as before. In the wake of collective action and state intervention, competition is not eliminated; it merely becomes more regulated.

A constructive effort has been made to maintain competition. An inherent weakness of competition lies in its dependence upon contract. So long as buyers want goods and sellers are willing to accept money, the liberty of contract promotes order. The greed for more money makes it advantageous to the producers of goods to agree among themselves to control production capacity, to regulate output and to stabilise prices. In such cases, the state is invoked to make the rivals compete rather than co-operate. Freedom of contract is encouraged along vertical lines (i.e., between buyer and seller) and has to be prohibited along horizontal lines (i.e., between buyer and buyer and between seller and seller).

3.3.6 Unemployment

Work satisfies many needs of the individual and the community. For the individual, work satisfies the need to exercise his faculties and to participate in the collective work of society. In addition, work also gives the individual a claim upon social products, enabling him to support himself and his family. In case of the community, work is necessary for survival and progress of civilisation.

In traditional societies, in which the productivity of agricultural labour is very low, virtually the entire population must be employed in farming. When productivity reaches a certain level, the demand for *primary goods* (i.e., agricultural products) drops in relation to the demand for other goods like clothing, shelter and manufactured products. The production of these *secondary goods* ultimately becomes organized in factories and expands dramatically. As the demand for manufactured goods grows and remains high, employment in the *secondary sector* also remains high. In addition, there is also a *tertiary sector* of employment comprising of services like teaching, administration, medical care, tourism and other similar pursuits that are not carried out in factories. In countries with high standard of living, the demand for products of the tertiary sector keeps increasing. As a result, employment in this sector increases more rapidly than in the primary or secondary sector.

The great shrinkage of employment in the primary sector is one of the most important phenomena of modern history. Men who abandon the soil have to change not only their means of livelihood but their residence and way of life. For a long time, the migration from a peasant culture involving millions of people, went mainly towards the factories. Although the output of the factories continued to increase, the same was not the case with factory employment. In the US, employment in the secondary sector peaked at about one-third of the labour force from 1920 to 1970 and since then, the expansion has been in the tertiary sector. Technological progress would have led to unemployment (or much reduced working hours), had it not been for the expansion of the tertiary sector. The same is more or less true in case of other developed countries.

The consumption of primary or agricultural goods eventually reaches a point of saturation. The consumption of manufactured goods passes through a phase of increase and then another phase of relative decline. It is the tertiary sector, however, that absorbs most of the manpower freed by technological progress in the other two sectors. The result is that employment as a whole does not decline over the long run because of technological progress. In the wealthy and technologically advanced countries, neither the size of the work force nor the number of working hours per week has shown any tendency to decline in the last fifty years. Of course, there have been economic crises (like depression and recession) giving rise to unemployment; but the unemployment created in this way was eventually absorbed. A dynamic economy requires that the labour force be mobile enough to move out of the sectors in which technological advances have reduced the need for manpower and into the sectors in which labour is in short supply. Such a migration, however, is inevitably accompanied by some degrees of unemployment or underemployment.

During the Great Depression of the 1903s, the rate of unemployment in the western capitalist countries reached very high levels. In the US, the rate of unemployment reached 25% of the labour force in 1933. In Sweden, unemployment reached 25% twice (in 1921 and 1931). Even in Great Britain, unemployment was above 15% during the period of Great Depression. Since the Second World War, unemployment rates in developed countries have remained far below such catastrophic levels, but they are quite high in most of the developing nations.

Even when the *overall* rate of unemployment is low, it remains a serious social problem. While some workers may be temporarily out of work (or passing from one job to another), others may remain unemployed for a long time. The US Bureau of Labour Statistics has studied the incidence of unemployment quite thoroughly. The figures of this Bureau show that women have more unemployment than men, young people more than adults, and persons without education or special skills more than educated persons.

Since the Second World War, the governments of many countries (especially the developing ones) have become committed to a programme of reducing unemployment and underemployment. The constitution of France explicitly charges the state with assuring full employment to its citizens. Similar goals have been set up with varying degrees in many other countries.

Even, the Charter of the United Nations makes full employment a major objective for its members.

Different governments have followed various policies in the pursuit of full employment. One general approach towards this end is to improve the supply of manpower; another is to alleviate the adverse effects of unemployment and underemployment. A third approach to full employment seeks to maintain the economic activities at a high level through fiscal policies. Finally, there is the method of economic planning where the government's Planning Commission sets targets for various sectors of the economy that are linked to forecasts of available manpower. This approach is based on the belief that the complex problems of unemployment and underemployment cannot be separated from other problems of economic and social developments. Moreover, all the developed countries and many developing ones as well try to soften the impact of unemployment through some forms of unemployment compensation. In some cases, governments seek to induce employers to retrain workers for new jobs rather than laying them off. The problem of unemployment is often a regional matter. When this is the case, attempts are made by governments to mitigate it through regional development programmes.

The efforts to deal with the *instability* of employment caused by technological and economic progress involve providing information on the state of the labour market and on the qualifications of those seeking work. This is done by government or private employment agencies in the hope of directing job seekers more efficiently to existing jobs or helping them to prepare for occupation in which manpower is likely to be needed. Most countries now have such agencies, designed to bring together the two sides of the labour market. Information provided by such employment agencies leads to guidance, and guidance leads to education, training and retraining. Many governments endeavour to provide or subsidise training programmes for those who could benefit from them.

The use of monetary and fiscal policies to keep the economy functioning at a high level of employment has been undertaken in many countries since the Great Depression. In periods of recession or of growing unemployment, the government may increase the aggregate demand by expanding the money supply or by increasing its own spending. This approach was quite successful in the US during the 1960s, when a major reduction in the income tax, incentives for business investments, and a large increase in federal spending for both war and non-war purposes brought the unemployment rate down to 3.5% (in 1969). In the years that followed, the US was faced with the problem of serious inflation. In order to solve the problem of inflation, efforts were made to stabilize prices; but these efforts directly led to rising unemployment. In contrast, the French government undertook national planning on a *broad scale* in the 1950s and 1960s and succeeded in maintaining rapid economic growth, full employment and stable currency. This shows that the problem of unemployment and underemployment cannot be divorced from other problems of economic and social developments. Thus, the employment policies will have to be oriented towards a context much broader than placement bureaus, labour mobility and fiscal policies.

3.4 SOCIETAL RESPONSIBLITIES

3.4.1 Man as a Social Animal

Man is a social animal by nature and necessity. He lives in society for his very existence and welfare. There is hardly any aspect of human life in which man does not feel the need of society. He cannot live in isolation since he is compelled to live in society both biologically and psychologically. The relationship between the individual and the society is neither contractual nor organic; it is something much more than that. It is a social need for man to live in the company of his fellow human beings in order to develop his personality and individuality. Human nature is such that it cannot be developed without the help of other human beings. There are several recorded instances of isolated children (the so-called "wolf-children") who, when they grew up, had nothing of human qualities. They imbibed only animal qualities like growling, walking like a four-footed animal and eating raw flesh.

It has been said that if nature impels man to lead a social life, necessity compels him to do so. Man has a variety of needs. In the first place, he has basic physiological needs since he cannot live without food, water, sleep, shelter, warmth and light. Secondly, man needs safety and he must defend himself against wild animals, extreme heat and cold, infectious diseases, etc.; so he needs others to stand by him. Thirdly, man has the need for love since everyone wishes to be loved by others. Fourthly, man has the need for social recognition. He wants that his distinctive qualities and special talents are recognised and respected by the society. Finally, man has the need for self-actualisation. Each individual has a self-image and realisation of this self-image or self-actualisation is a psychological necessity. These diverse physical, mental and spiritual needs of man can be satisfied only within the framework of society, and it is for the society to arrange the required framework to satisfy the basic needs of man. Thus, the society has certain responsibilities for the well-being and development of its members. Some of these responsibilities are:

(1) Public health services.
(2) Public education.

(3) Welfare of backward classes.

(4) Family and child welfare.

(5) Welfare of vulnerable sections of the population.

(6) Full employment.

(7) Protection from natural disasters and external threats.

(8) Housing and clean environment.

(9) Public transportation.

(10) Facilities for communication.

(11) Recreational facilities.

This list is merely indicative and not exhaustive.

3.4.2 Public Health Services

Public health has been defined as the art and science of preventing disease, prolonging life and promoting physical and mental efficiency through organised community effort. In primitive communities, the main strategy for defence against sickness was to isolate or even to destroy the sufferer. As the knowledge of sources and modes of infection increased, governments tried to control the implicated environmental factors such as public water supply, milk and other foods, human wastes, insects, various forms of environmental pollutions, etc. Attention has also been paid to housing and special diseases like cancer, diabetes, venereal diseases, cardio-vascular diseases and mental illness. Thus the public health emphasis has gradually changed from an impersonal approach (in primitive societies) through environmental controls to a more personal approach through preventive medicine (in modern societies). The full range of health services (promotive, preventive, curative and rehabilitative) is now inseparable. As a result, the emphasis in *advanced* countries is on developing methods to integrate these sectors into comprehensive health service systems aimed at both the individual and the community.

In *developing* countries in Asia, Africa and Latin America, the difficulties of providing health services involve several interrelated problems which arise from the nature of the diseases prominent in the region, inadequate and improperly distributed resources, design of inadequate health service systems and improper education of health personnel working in those systems. Government and society have several responsibilities to improve public health services. The principal causes of sickness and death among small children in the developing countries are diarrhoea, respiratory infections and malnutrition, all of which are diseases intimately related to culture, custom and economic status of the country involved. Another factor that contributes to health problem in developing countries is the large family size. Malnutrition occurs most often in children born into large and poorly spaced families. The resulting high death rate among small children in such families often reinforces the tendency to have more children and, the vicious cycle continues. Money is a crucial factor in health care. Also suitable mental hospitals, child guidance and marriage guidance clinics and schemes for the care of alcoholics and drug addicts are essential. There should be significant developments in the treatment of maladjusted members of society.

The fragmentation of earlier health service organisations (such as single-disease-oriented programmes and the separation of curative and preventive services) is now giving way to more comprehensive organisations. Health promotion, disease prevention, curving of the ill and rehabilitation are brought together into one network of integrated services that reach the community level.

Decisions of great complexity are involved in allocating limited resources to provide health services to a large number of people. In order to achieve optimum results, there should be an increasing emphasis on the health planning process and on the design of more effective public health service systems. An important aspect of national health planning should be close co-ordination between planning, budgeting, implementing and evaluating of health-care programmes. No public health service can be fully effective unless the concerned government agency and the society pay close attention to this aspect.

3.4.3 Public Education

The essential features of a system of public education are the requirement of compulsory attendance, free tuition, provision of books and educational opportunity for all. In most countries, very few of these ideals have been realised in practice. Free education, for example, is generally limited to primary education in most countries. The additional proposal that free tuition should be supplemented by scholarships for maintenance of the students is still in the discussion stage even in the leading countries. Today there is a general tendency all over the world for compulsory education up to the age of fourteen. While the principle of compulsory education up to the age of fourteen has generally been accepted, the enforcement of compulsory attendance still lags far behind. This is partly due to poor economic conditions and inadequate number of schools in many countries and partly as a result of the inertia of parents who themselves have had no education.

The concept of public education, however, should not be limited to the provision of schools only. It should include the provision of all those activities and organisations that enable the students to derive the fullest benefits from the school work. These generally include provision for medical inspection, medical treatment, meals in school (with or without charge), school libraries, transportation (with or without charge), school camps and excursions and, above all, adequate playgrounds and athletic facilities. To these should be added the agencies for vocational guidance and placement. It may be mentioned, that *complete realisation* of this combination of education and social services is not to be found in any one system of public education anywhere in the world. The reason is that the above conception of the scope of public education is of very recent origin. Although extracurricular activities such as various clubs, Boy Scouts, Girl Guides, NCC, etc., are found everywhere, these have not yet been incorporated in public school systems under the control of public authorities. It is now responsibility of the society to see that its members get adequate public education.

3.4.4 Welfare of Backward Classes

A social welfare policy may be defined as the strategy of action indicating the means and methods adopted to implement the social welfare services which include the following:

(1) Welfare of backward classes.
(2) Welfare of vulnerable sections of the society.
(3) Family and child welfare.
(4) Correctional services.

These services include programmes which are intended to cater to the needs of persons and groups who, by reasons of social, economic or physical handicaps, are unable to avail of (or traditionally denied) the amenities and services provided by the community. In other words, social welfare services offer services to those sections of the society who need special care. The vulnerable sections of population, on the other hand, include persons belonging to backward classes, scheduled castes, scheduled tribes, children, youth, women, slum dwellers, physically or mentally handicapped, women under moral danger, juvenile offenders, beggars, prisoners, etc.

In spite of some remarkable advancements, the overall progress is not commensurate with the expenditure. There is a need to bring about a change in the existing organisational set-up for the implementation of social welfare policy in order to achieve integration and co-ordination among various social welfare services for the weaker sections. And the society has a tremendous responsibility to achieve the objectives.

3.4.5 Family and Child Welfare

Several programmes for family and child welfare have been formulated and implemented in India since independence, but an explicit national policy for family and child welfare has not yet been evolved. The major emphasis at present is on the integrated approach to provide welfare services to family and children. At present, Family and Child Welfare Projects are the important programmes directed towards the welfare of women and children especially in rural areas. The Family and Child Welfare Projects normally cater to the children in the age group of 0-16 years and greater attention is paid to the children in the age group of 0-6 years. The Family and Child Welfare Projects are implemented at the block level. The responsibility of fulfilling the statutory obligations for family and child welfare under various central and state legislations rests mainly with the government agencies. The voluntary social welfare organisations deal mainly with non-statutory services for the welfare of women, children, the handicapped and other vulnerable groups. Thus, the voluntary social welfare organisations share a major responsibility in the implementation of social welfare policy. Voluntary organisations in the society can play a crucial role in the field of family and child welfare due to their vast and pioneering experience in social welfare and their humane approach (in contrast to the bureaucratic approach by government agencies). The society has thus a pivotal role to play for family and child welfare.

3.5 SOCIAL INSTITUTIONS

3.5.1 Concept of Social Institutions

Social institutions are organised ways to meet the basic needs of a society. Institutions usually involve norms that guide social interactions and thus, reducing the likelihood of random or unpredictable behaviour by the members. A social institution can be defined as a cluster of norms that guide social interactions towards the fulfillment of one or more of the basic needs of a society. When these norms are accepted, persons interact socially in predictable ways that lead to fulfillment of societal needs. Education, religion, family, political system and the economic system are regarded as the basic social institutions in most societies. The concept of social institution can be broken down into three basic elements:

(1) A social institution helps to fulfil one or more of the basic needs of a society.
(2) Each social institution lays down a set of norms.
(3) Social interaction is guided by the norms of a social institution and these norms are expressed in the roles associated with various social positions.

The first element explains the purpose or function of a social institution. Each society has certain fundamental needs, which must be met if the society has to survive. Social institutions help meet these needs. Economy meets the need for efficient production and exchange of goods and services, while the family meets the need for reproduction and socialisation of new social members. The second element involves a cluster of norms. These norms specify certain procedures that should be followed like paying interest on borrowed money and buying goods at a certain rate. The third element included in the concept of a social institution is that interaction among persons is greatly influenced by social positions and accompanying roles associated with an institution. Each institution has a number of social positions like mother and father in the family, block and district representatives in the political institution and students and teachers in the educational institution.

A social institution usually exhibits the following characteristics:

(1) It emerges through unplanned development.
(2) It changes slowly.
(3) It is related to other institutions.
(4) It assumes different forms in different societies.

The development of a social institution is the result of a gradual evolutionary process. In case of a family, the members of a society do not consciously agree that their family institution has to take a particular form. Rather it emerges over a period of time in accordance with several other social changes. Similarly, changing sex roles in the institution of a family in modern Indian societies have significantly influenced the institution of economy. Employment of women on a massive scale has contributed to many changes in the family institution. The norms and forms of social institutions in one society may bear little resemblance to the same institutions in another society. For example, political, economic, religious, educational and family institutions in the US are quite different from that of China.

3.5.2 Functions of Social Institutions

The term "institution" in its social usage implies a way or thought or action which is embedded in the habits or customs of a group of people. The range of institutions is as wide as the interests of mankind. Any *informal* body of usage (such as common law, higher education, moral code, etc.) is an institution in the sense that it lends sanctions, imposes taboos and lords it over some human concerns. Any *formal* organisation like the government, the church, the university and the trade union imposes commands, lays out penalties and exercises authority over its members. All these are examples of social institutions. They may be rigid or flexible in their structures, strict or lenient in their demands, but all of them constitute standards of conformity from which an individual member may depart only at his peril.

Our culture is a synthesis or a collection of institutions. Each of these institutions has its own domain and its distinctive office. The function of each social institution is to set a pattern of behaviour and to fix a zone of tolerance for various activities related to the institution. Etiquette, for example, decrees the rituals which must be observed in all polite social intercourse. Education, on the other hand, provides civilizing exposures through which the potential capacities of individuals are developed into the abilities for performance, appreciation and enjoyment. The institution of marriage gives propriety to the sex union, bestows regularity upon procreation, establishes the structure of the family and creates a balance between personal ambition and social stability. A number of social institutions may combine or compete to impress character upon the mass of human endeavour and to give direction to it. Thus, the state claims primary obedience and imposes some order upon the activities of mankind. The institution of law determines the outmost limits of acceptable actions by punishing offenders and settling disputes. The community is made up of such overlapping provinces of various social institutions. It is the social institution in its role of organiser which makes this world a social world.

3.5.3 Origin and Development of Social Institutions

It is almost impossible to discover a legitimate origin for such an organic and complex entity as a social institution. Its origin may lie in an accidental, arbitrary or a conscious action. A man (savage or civilised) strikes a spark from flint, makes an image from mud or brews a concoction. The act is repeated and then multiplied; ideas, sanctions and habits from the existing culture get attached; and gradually there develops a ritual of fire, a ceremonial for appeasing gods, or a cult of healing. In all societies, however forward or backward, the roots of even the most elementary arrangements like barter, burial, worship, work life and sex union run far back into the unknown past and embody the knowledge, ignorance, hope and fear of people. In fact, a social institution has no origin apart from its development, since an institution is an aspect of a continuous social process. A social institution emerges form the impact of novel circumstances upon ancient custom and it is transformed into a different group of usages by cultural change. In the growth of a social institution, the usual may give way to the unusual so gradually as to be almost unnoticed. As an institution develops within a culture, it responds to changes in the prevailing sense and reason. The public regulation of business consistently reflect the prevailing thinking on the relation between the state and the industry. Similarly, the pages of law reports reveal the ingenuity with which the same old rules and standards are reinterpreted to serve the changing notions of social necessity. In this continuous process of adaptation by an institution to the prevailing intellectual environment, an active role is assumed by the *common sense*, i.e., the body of ideas taken for granted by a society during a particular period. Because commonsense determines the climate of opinion within which all other institutions must operate, it is the dominant institution in a society.

In an even broader way, an institution is accommodated to the folkways of culture in a society. As circumstances impel and common opinions in a society change, an institution held in high esteem earlier (e.g., piracy) may fall from grace, while another under taboo (such as birth control) may at first win tolerance and later general acceptance. As one social system passes into another and the values of life change, one social institution gives way to another that is better adapted to the times. An institution that survives, such as matrimony, has to respond to cultural changes and adapt to them. In the social process, the life of an institution depends upon its capacity for adaptation to changing social conditions.

The same process of development applies to an institution introduced from an alien society. The act of borrowing a social institution from another society merely gives the opportunity for its modification to suit the needs of the adopting society. When Russia appropriated the *Industrial Revolution*, it stripped away the enveloping business arrangement and converted it into an instrument to serve as a national social economic system. The act of transplantation of a social institution into another society may at first retard its growth, but eventually it is likely to promote it.

3.5.4 Functioning of Social Institutions

The very flexibility of a social institution makes it a creature of social stress and strain. In a stable or a slowly-changing society, an institution fits rather neatly into the cultural pattern of the society. On the other hand, if social changes bring disorder, the structure of a social institution may be compromised. A social institution may even fall into the hands of "enemies" and be used to defeat its proclaimed purpose. Thus a community of ascetics often develops into a wealthy religious establishment, a political party dedicated to personal freedom becomes the champion of vested wealth and a philosophy contrived to liberate thought remains to enslave it. Those who contrive rules and formulae cannot control the uses to which they are put. In the course of time, therefore, the function of a social institution may be compromised or even lost in its establishment. The spirit may become the letter and the vision may be lost in a ritual of conformity.

If a social institution becomes formal, a greater hazard to its integrity is found in its organisation and its personnel. Thus, a need for law and order finds an expression in a government, the demand for justice in a legal system and the desire for worship in a church. When the government, legal system and church are formally established, various groups become interested in their structure and offices, their procedures and emoluments and their ceremonials and traditions. A host of officials come into being, who are mainly interested in the maintenance of the establishment to which they are committed. These officials have their own preferences and prejudices and they are not immune to consideration of their own prestige and position. As an institution becomes more formal and more rigid, the good of the institution rather than its proclaimed purpose tends to become dominant. When this happens, the lines of activity of the formalised institution may be frozen into rigidity.

As long as a social institution remains vital, men accommodate their actions to its detailed arrangements with little worry about its inherent nature or its cosmic purpose. When it begins to give way or it is seriously challenged, compelling arguments are set forth to justify the existence of an institution. For example, the institution of *capitalism* was never created by design; but now that it is already there, contemporary scholars have intellectualised it into a purposive and self-regulating instrument of general welfare. So long as people are able to do as their fathers and grandfathers did, they manifest little curiosity about the arrangements under which they live and work. So long as the procedure of institution is unquestioned, people are little aware of the conventions and values which give rise even to outstanding achievements of an institution.

3.5.5 Organic Nature of Social Institutions

Institutional development always drives a fault line between current fact and prevailing opinion. Men meet new events with the wisdom they already possess. That wisdom usually belongs to the past since it is a product of the experience of a by-gone era. As new social institutions emerge from the old, men persist dealing with the unfamiliar in their old and familiar ways. Thus a social institution, like a living thing that it is, has a tangled identity. It cannot be shown in perspective or revealed in detail by the logical method of exclusion and inclusion. Each institution holds within itself elements drawn from the modern era of information technology, the rational universe of the eighteenth century and the folkways of some far off centuries. It holds many unknown possibilities, which a suitable occasion may bring to life.

It may be concluded that a social institution is an imperfect agent of an order and purpose in a developing culture. It is created by both intent and chance. A social institution imposes its pattern of conduct upon the activities of men and its compulsion upon the course of unanticipated events. A social institution may, like any other creation of man, be taken into bondage by the power that it was designed to control. An institution is a folkway, always new and yet ever old, directive and yet responsive, a creature of means and also a master of ends.

MODEL QUESTIONS

(Essay/Long Type)

3.1 Define and explain "social mobility".

3.2 Explain in detail the vertical mobility.

3.3 Write in brief the consequences of mobility.

3.4 What is understood by "income"? Explain in brief.

3.5 Define and explain "national income".

3.6 Write a note on income distribution.

3.7 What are the causes of inequality of income? Explain in brief.

3.8 What are the consequences of inequality of income? Explain.

3.9 Explain the causes behind social tensions.

3.10 "Class struggle creates social tensions" – Justify the statement.

3.11 Explain "political corruption".

3.12 Define crime. Give reasons behind crime. Suggest steps to control crime.

3.13 Write a note on "Competition".

3.14 What are the major causes behind unemployment? Suggest some measures to reduce the rate of unemployment.

3.15 List the societal responsibilities and explain each in very brief.

3.16 "Public health services, Public education, Welfare of backward classes and Family and child welfare are the major societal responsibilities" – Explain to justify the statement.

3.17 What is understood by "social institutions"? Explain.

3.18 Write in brief the functions of social institutions.

(Objective/Short Type)

3.19 What is a horizontal mobility?

3.20 Give an example of a vertical mobility.

3.21 Define real income.

3.22 List the possible measures to reduce inequality of income.

3.23 Write the Pareto's law mathematically.

3.24 Draw the Lorenz Curve for equal distribution of income.

3.25 The strategies to combat crime are:

    (a) Punitive   (b) Therapeutic

    (c) Preventive   (d) All the three.

3.26 Match the following:

    (A) Communist Manifesto   (a) Nigeria

    (B) NCC   (b) Karl Marx

(C) Corrupt country   (c) Mahatma Gandhi

(D) Tertiary sector   (d) India

3.27 Write true or false :

The mobility is said to be intergenerational when an individual moves vertically upward or downward within his own lifetime.

3.28 Write true or false :

Opportunities for mobility are greatly enhanced during a recession or a depression.

3.29 Write true or false :

The total income of an individual is the total money value of the services received by him from all sources, including his own activity.

3.30 Write true or false :

Social institutions are never the organized ways to meet the basic needs of society.

3.31 Name the odd ones :

(a) Health promotion   (b) Family welfare

(c) Class conflict   (d) Rehabilitation.

3.32 Fill up the blanks :

The five main causes of social tensions are _____, _____, _____, _____ and _____.

Development Processes

4.1 DEVELOPMENT OF TRADITIONAL SOCIETY

4.1.1 Traditional Society

The concept of development presents many difficulties when we try to define it in precise terms. The difficulties arise due to complexity of the process of development operating in many different cultures and under conflicting political ideologies. As a result, every analyst brings a new dimension to the concept of development. The main characteristics, of traditional society in India as it existed almost upto the end of eighteenth century are :

(1) Pre-machine technology.
(2) Face-to-face communication.
(3) Lack of belief in progress.

Pre-Machine Technology

The most important characteristic of this society was that it developed within the framework of limited pre-machine technology. This traditional society was incapable of continuous flow of innovations which were necessary for sustained growth. The pre-machine technology was characterised by several important features. In the pre-machine technology, the techniques of production in any field ran into a fixed grove with the result that any particular craft or skill remained virtually the same for centuries. For example, the methods of pottery-making or basket-weaving remained more or less the same for centuries. Another feature of the pre-machine technology was the high degree of animal and human energy used in it. That is the main reason why it is called pre-machine technology. This technology functioned in a society which lacked the basic ingredients for any fundamental changes in the social structure. The caste system led to rigid modes of thinking and patterns of life. The commercial castes were trapped into a commercial way of life and they had no incentive to bring technological knowledge into the society from outside or generate it themselves. As a result, the traditional Indian society did not develop the mental climate which could have produced an industrial revolution as was in Europe.

Face-To-Face Communication

Traditional societies are generally characterised by the pattern of their communications system, termed as "face-to-face". This resulted from the lack of geographic mobility and the absence of infrastructure for communications. The cellular character of the Hindu society with its intricate network of the caste system prevented communication between members of sub-castes. The caste system did not allow vertical social mobility and imprisoned individuals into fairly well-knit groups. The communication pattern in the traditional Hindu society also determined the attitude of the masses towards the rulers, the legitimacy of whose rule was taken for granted. As a result, there was little participation in the process of government. Kings ruled with an aura of divinity and the masses were merely "subjects", not "citizens". Both the ruling elites and the religious elites were concerned primarily with the functioning within the existing social system.

Lack of Belief in Progress

In all traditional societies, the political and administrative institutions were marked by a low degree of functional specialisation. The idea of progress was not totally absent in traditional societies since the idea of spiritual progress (in terms of individual salvation or about cosmic evolution of man) was fairly common. What the society lacked, however, was a belief in the possibility of goal-oriented secular changes through the organised activities of men.

4.1.2 The Process of Development

Development is a *total* transformation of society and a movement in a consciously-chosen direction. There is no straight and linear progress from traditional to modern society, and neither is this transition smooth. Development involves *change of outlook*, a belief in the possibility of progress and the acceptance of objective criteria and standards for constant improvement in all aspects of individual and social life. The essence of development lies in the awareness that men have the power to direct their lives and transform the society. Borrowing technology from advanced countries might indicate a desire to be developed, but it is not the essence of development. In fact, development is an attitude of mind; it is not to adopt, but to participate in; not to have, but to do, to be and to keep becoming. In brief, development is a complex one and involves several aspects:

(1) Political development.
(2) Social development.
(3) Economic development.
(4) Intellectual development.

Political Development

The first requirement of political development is a high degree of role of specialisation (and differentiation) of political institutions and the growth of communication media. Functional differentiation is represented by political parties, trade unions, religious organisations, pressure groups and other similar organisations. In the second place, there is greater emphasis on rational, scientific and secular techniques for decision making. The developed system acquires an impersonal character in which the law becomes more important than the whims of men in power. The bureaucracy develops into a highly specialised and reliable machine for tackling the problems facing the society, in a systematic and scientific manner. To achieve this goal, the bureaucracy has to be manned by a class of highly trained and professionally oriented civil servants. In the third place, there is a great deal of mass mobilisation by the people. It is important, however, not to identify political development with a liberal democratic society since development cuts across political ideologies.

Social Development

When a traditional society is developed, there is a marked shifting of population from rural areas to urban centers under the impact of industrialisation, which is one of the main agents of development. In addition to migration to cities and towns, the growth of technology gradually reduces the percentage of the population engaged in agriculture. Social development brings about a marked change in the outlook and behaviour of social groups which are characterised by the functions they perform rather than by their caste, language and other such factors. With the growth of social development, the individual finds himself in a wider world of freedom in which there are fewer restrictions on his ability to take decision affecting his life. He is free to choose his own career rather than have it determined on the basis of his caste.

In a developed society, the communication system links up different sectors and sections of life, supplementing the face-to-face communication pattern of the traditional society. Of course, the vast communication-network calls for a body of professional communicators such as journalists, writers, public speakers, etc. Another characteristic feature of social development is that it puts a premium on the acquisition of skills and achievement of desirable results rather than on birth or other social factors like caste.

Economic Development

Economic development also known as "economic growth" covers many aspects of social life. In the first place, it involves the systematic application of science and technology to the processes of production and distribution of goods and services. Secondly, it compels increasing use of inanimate sources of energy in contrast to the use of human or animal energy in traditional societies. This change in the pattern of energy consumption can only be sustained by a revolution in the consumption patterns of the masses demanding diversification of production in response to varied consumer needs. Diversified consumer needs lead to a high degree of specialisation in production techniques and labour skills. Rationality in economic decisions (in determining the location of industry, for example) results in increased mobility of labour and emergence of a vast variety of market processes.

Economic development of a society is impossible without bringing into existence specialised human skills. Economic development, like political development, gives rise to functional differentiation and to various financial institutions which feed the economic system. Economic development can not be self-sustaining without setting up financial and other institutions by which the professional, managerial and technical skills are imparted. As the development proceeds, there is a gradual shift from the extractive industries to the manufacture and provision of consumer goods. The linking up of natural resources with the location of industries and the efficient operation of distribution channels require a well-developed transport system. Finally, there is an emergence of a labour force conscious of the important role it is called upon to play in a modern economic system. Another important aspect of economic development is *advertising* which has brought about significant and far-reaching changes in modern life. Due to revolution in communication media, advertising is able to link the ever-increasing number of consumers with a variety of goods and services. Advertising has encouraged launching of new brands and made the consumer conscious of his freedom of choice. In fact, advertising is one of the most reliable indices of development.

Intellectual Development

Development cannot be sustained for long in any society without a corresponding and self-sustaining intellectual development characterized by constantly increasing knowledge. This involves the existence of adequate number of fact-finding and data-processing agencies, statistical units, research and development laboratories, universities and similar institutions. Intellectual development implies the existence of intellectual elites who play a key role in sustaining the growth of technology. Intellectual development leads, in all political systems, to greater emphasis on secularism and on secularisation of the process of government and bureaucracy. It also leads to an increasing emphasis on strengthening the material basis of life.

4.1.3 Impediments to Development

There are several impediments to development of a traditional society. Some of the obstacles are :
  (1) Lack of skills.
  (2) Rigid administrative system.
  (3) Impatience for rapid development.
  (4) Passion for quantitative expansion.
  (5) Premature politicisation.
  (6) Strain on law and order resources.
  (7) Rapid growth of population.

(1) Lack of Skills

The developing countries are usually weak in the skills required for development. The real problem in training personnel for development programmes lies not in imparting information to them, but in helping them to develop the required skills. It is necessary, to give field workers and administrators more freedom to experiment and to try new approaches; but this is precisely where the Indian programmes suffer.

(2) Rigid Administrative System

In India and in many other Commonwealth countries, the administrative system inherited from the British rule leaves little room for freedom to experiment. The inherited bureaucracy with its outmoded procedures of work and personal attitudes, inadequate delegation at all levels, too formal supervision of field workers and poor morale provide a major impediment.

(3) Impatience for Rapid Development

It arises from the belief that a country must embark on all areas of development at one time. This has led, among other things, to symbolic expenditure on big projects to convince the masses and the outside world of the country's determination to become a modern nation in the shortest possible period of time. Many poor countries have spent huge money on nuclear research even though the basic amenities of life remain unprovided for a high percentage of their population.

(4) Passion for Quantitative Expansion

Another obstacle arises from the passion for a rapid quantitative expansion without attention to quality. Apart from community development, education has very rapidly expanded in India since independence and new universities and colleges have mushroomed under local pressure. The result of this expansion has been pumping into the society a vast army of unemployed graduates.

(5) Premature Politicisation

The political leadership in developing countries has a marked tendency to politicize the mass prematurely. The large number of students and unemployed youths, often recruited by various political parties, contribute to the restlessness of the political process.

(6) Strain on Law and Order Resources

Politicisation of the mass results in considerable strain on the law and order resources of the state. The leadership in India has done very little since independence for rehabilitating the police in the popular mind as protectors of the law. Attitude formed in the popular mind towards police in the era of our freedom struggle has not yet died but has produced a certain ambivalence towards the police. As a result, investment in improvement and strengthening of the police department has appeared to our leadership as being in some way contrary to the spirit of democratic welfare.

(7) Rapid Growth of Population

Rapidly growing population is one of the major impediments to the development of a traditional society. Rapid population growth usually results from the improvement in the general conditions of the mass, better health-care facility and decline in morality. A high rate of population growth offsets the economic growth of a country. This leads to frustration, social tension and mass violence.

4.2 THE PROCESS OF DEVELOPMENT

4.2.1 What is Under-Development?

It is not an easy task to define under-development or an under-developed country. According to the "*United Nations Experts Committee,*" an under-developed country is one whose per capita real income is low when compared with the per capita real income of the US, Canada, Australia and Western Europe. According to *Prof. Ragnar Nurkse,* "the under-developed countries are those which, compared with the advanced countries, are under-equipped with capital in relation to their population and natural resources". Even this definition is not fully satisfactory. The *Indian Planning Commission* has defined an under-developed country as one "which is characterised by the co-existence, in greater or lesser degree, of unutilized or under-utilised manpower on the one hand and of unexploited natural resources on the other". According to *Colin Clark,* who was one of the pioneers in the studies of under-developed economies, "economic development consists in the progressive enlargement of tertiary occupations in the economy". According to this definition, under-developed economies are those in which the primary occupations predominate. In current literature, all countries with low per capita income are generally classified as under-developed. In general, all countries with per capita income less than 10% of that of the US (31,910 US dollars) may be regarded as under-developed countries. India, with a per capita income of 2,230 US dollars, is one of the most under-developed countries in the world.

4.2.2 Characteristics of Under-Development

Some of the basic common characteristics are discussed below:

(1) Low Per Capita Income

The per capita income in under-developed countries is low compared to that of developed ccuntries. For example, the per capita income of India, Pakistan and many other developing countries of Asia and Africa is less than 2,500 US dollars, while that of the US and Japan is more than 25,000 US dollars. Low per capita income is such an outstanding feature of an under-developed economy that it can be used as a yardstick to measure the level of development of a country. The per capita income of some under-developed countries is shown in Table 4.2.2.1, while that of some developed countries is given in Table 4.2.2.2.

(2) Deficiency of Capital

The insufficient amount of capital, an important characteristic of all under-developed economies, is often called *capital-poor* economy. In most under-developed countries, the rate of investment is only 5-8% of the national income, while it is 15-18% in the US, Canada and Western European countries. In an under-developed economy, the low level of per capita income limits the size of the market for goods and services which, in turn, weakens the inducement to invest. The low level of investment also arises as a result of the lack of *dynamic entrepreneurship,* which may be regarded as the focal point of economic development. The root of capital deficiency is the *shortage of savings.* The level of per capita income being quite low in

Table 4.2.2.1   Population and per capita income of some under-developed countries
(Ref: Manorama Year Book, 2002)

Name of the Country	Population (million)	Per capita income (US dollar)
Congo	3.1	540
Malawi	11.6	570
Burundi	6.5	570
Afghanistan	22.5	800
Rwanda	7.9	880
Bhutan	2.1	1,260
Nepal	23.6	1,280
Bangladesh	140.4	1,530
Pakistan	145.0	1,860
India	1,025.0	2,230

Table 4.2.2.2   Population and per capita income of some developed countries
(Ref: Manorama Year Book, 2002)

Name of the Country	Population (Million)	Per capita income (US dollar)
Singapore	4.1	22,310
Hong Kong	7.0	22,570
Germany	82.0	23,510
Australia	19.3	23,850
Austria	8.1	24,600
Japan	127.3	25,170
Canada	31.0	25,440
Belgium	10.3	25,710
Switzerland	7.2	28,760
United States of America	285.9	31,910

a developing economy, most of it is spent in satisfying the bare necessities of life (like food, clothing and shelter), leaving very little for capital accumulation.

(3) Excessive Dependence on Agriculture

It is a well-known fact that most under-developed countries are predominantly agricultural. A great majority of their population (usually between 70 and 90%) are engaged in agriculture and allied occupations. This excessive dependence on agriculture is due to the fact that non-agricultural occupations do not grow at a rate commensurate with the population growth due to lack of sufficient investments in areas other than agriculture. The labour to land ratio being high in an under-developed economy, agricultural holdings become subdivided into very small plots, which do not permit the use of modern mechanical methods of production.

(4) Rapid Growth of Population

Although there is a diversity among various under-developed economies with respect to their population, they all share a common feature, viz., a high rate of population growth. This rate has been rising still more in recent years in many under-developed countries due to advances in medical sciences, which have greatly reduced child mortality and death due to epidemics. While the death rate has fallen phenomenally, the birth rate has not yet shown significant decline. The common trend of rapid population growth in under-developed economies is highly important. Its great relevance lies in the fact that it frustrates all attempts at development since the increased output is swallowed up by the increased population and the country continues to remain poor. The size of Indian economy, is the fifth largest in the world (after the US, Russia, China and Germany) from the point of view of the Gross Domestic Production (GDP). On the other hand, our population is so large that if we consider the per capita income, India lies among 10 or 15 poorest countries of the world.

(5) Large-Scale Unempolyment

An important consequence of the high rate of population growth in an under-developed economy without a corresponding high rate of economic growth is that there is a large-scale unemployment in urban areas and under-employment in rural areas. In India, the rate of annual growth of population is about 17 million. We are not able to create even a fraction of 17 million new jobs each year with the result that the absolute number of unemployed (or underemployed) persons goes on increasing every year. It means that more people have to remain engaged in agriculture than needed. It is obvious that the addition of such persons does not contribute much to the productivity of the land.

(6) Under-Utilisation of Resources

Various natural resources in an under-developed economy are either unutilised or under-utilised. Generally under-developed countries are not deficient in land, water, mineral or forest resources and these natural resources remain merely potential resources. These countries are not able to utlise their natural resources fully or properly due to a variety of reasons such as lack of capital, lack of proper equipment, inaccessibility to natural resources, primitive technology, small size of market, etc.

(7) Foreign Trade Orientation

It is a common observation that most under-developed economies are foreign-trade oriented. The under-developed countries export their raw materials to foreign countries instead of utilising them at home and import manufactured goods instead of making them in the country itself. Excessive dependence on exports makes the economies unstable and adversely affects the terms of trade. The tendency to import manufactured goods is also high in such countries.

(8) Low Levels of Skills

The under-developed countries employ primitive methods of production and utilise inferior technology. There is also a terrible dearth of skilled personnel. Poor techniques and low levels of skill result in inefficient and insufficient production, which leads to poverty. The economic backwardness or poverty of under-developed economies manifests itself in lower efficiency, illiteracy, ignorance and lack of entrepreneurship.

4.2.3 Stages of Development

Prof. Rostrow, an eminent economic historian and a specialist on economic development, has divided the historical process of development or economic growth into the following five stages :
   (1) The traditional society.
   (2) Pre-conditions for take-off.
   (3) The take-off period.
   (4) The drive to maturity.
   (5) The age of high mass consumption.

(1) The Traditional Society

In a traditional society, modern science and technology are either not available or, even if they are available, they are not being systematically applied, though there may be *ad hoc* applications of innovations. In a traditional society, agricultural production can be increased due to an increase in the acreage of land under cultivation and the composition of domestic and foreign trade may also change. But the distinguishing feature of the traditional society is that there exists a ceiling on the level of the attainable per capita output. A large proportion of productive resources in a traditional society is devoted to agriculture.

(2) Pre-conditions for Take-Off

This stage may require a long period, a century or even more, during which the pre-conditions for take-off are established. These conditions mainly consist of fundamental changes in the social, political and economic fields. Some of the conditions are :

(1) Change of attitude towards science and technology, risk taking, earning a profit, etc.
(2) Adaptability of the labour force.
(3) Political sovereignity.
(4) Development of financial institutions.
(5) Development of infrastructures such as roads, railways, communications power etc..

(3) The Take-Off Period

This is the crucial stage of development and covers a relatively brief period of two to three decades. During the take-off period, the economy of an under-developed country transforms itself in such a way that the subsequent economic growth takes place more or less automatically. The "take-off period" is defined as the interval during which the rate of investment increases in such a way that the per capita real output rises significantly. This initial increase in output carries with it the radical changes in the techniques of production and the disposition of income, which perpetuate the new scale of investment and the rising trend in per capita output. The term "take-off" implies three things. In the first place, the proportion of investment in relation to the national income must rise by 12-15%, thus outstripping the likely growth rate of population. Secondly, the take-off period must be relatively short, so that it can show the characteristics of an economic revolution. Finally, the take-off period must culminate in self-generating and self-sustaining economic growth.

(4) The Drive to Maturity

This stage consists of a long period of self-sustaining and self-propelling economic growth. During this stage, the rate of savings and investments is of su ·h magnitude that the economic development occurs almost automatically. As the economy matures, the overall capital per head increases considerably and that, in turn, changes the structure of the economy itself. The initial key industries, which sparked the take-off stage, now experience retardation as the law of diminishing returns comes into play. But a high rate of growth is maintained by a succession of new rapidly-growing sectors. During this stage of development, the proportion of population engaged in agriculture declines and, consequently, the structure of the developing country's foreign trade undergoes a radical change.

(5) The Age of High Mass Consumption

During this final stage of economic development, the per capita real income of the country is such that a large number of people can afford to buy the basic needs like food, clothing and shelter. There is a tendency for the leading sectors to shift towards durable consumer goods. There is also a steep rise in the proportion of population engaged in tertiary occupations. The present economies of the US, UK, Germany and Japan represent this stage of economic development. India launched her economic development programme in the form of *First Five Year Plan* from 1951-52. The first two Five Year Plans were mainly concerned with the preparation for accelerated development of basic industries, means of transport, facilities for technical training and development of agriculture. With the *Third Five Year Plan,* India entered a period of self-sustained economic growth, which has continued in subsequent Five Year Plans. The country has made tremendous progress in many areas such as atomic energy, space technology, food production, etc. But we are still in the *take-off stage* of economic development.

4.3 PARAMETERS FOR DEVELOPMENT

4.3.1 Economic Parameters

Economic development, opposite to under-development, implies that a country has a highly developed economic system. This also means that the country has a highly developed banking and credit systems and an adequate development of transport and communication systems. In addition to these, widespread social, institutional and organisational changes are also implied by the process of economic development. In other words, development means transformation of a predominantly agricultural economy into a highly industrialised one, bringing about a significant rise in the national and the per capita income, which is accompanied by a high rate of capital formation and a high degree of technological development.

The essence of economic development, commonly known as "development", is the growth of per capita output or per capita income. The process of development is a highly complex phenomenon and it is influenced by a variety of political, social and cultural forces. As observed by *Prof. Nurkse,* "Economic development has much to do with human endowments, social attitudes, political conditions and historical accidents. Capital is a necessary but not a sufficient condition of progress". The supply of natural resources, the growth of scientific and technical knowledge and several other factors also have a strong bearing on the process of economic growth. From the standpoint of economic analysis, the most important factors determining the rate of economic development are :

(1) The rate of capital formation.

(2) Capital-output ratio.

(3) The rate of population growth.

The Rate of Capital Formation

The crux of the problem of economic development in an under-developed economy lies in the rapid expansion of its capital investments. This is necessary so that the rate of the growth of output exceeds the rate of population growth. Only with such a high rate of capital investment will the standard of living begin to rise in an under-developed country. No economic development is possible without irrigation systems, production of agricultural tools and implements, land reclamation and construction of dams, bridges, factories, roads, railways, airports, ships and harbours. All these are the *"produced means of further production"*, associated with high levels of productivity. There is no doubt that the insufficiency of capital accumulation is the most serious limiting factor for the economic growth of a developing country. According to many economists, capital occupies the central and most strategic position in the process of economic development. However, the process of building up the necessary stock of capital equipment requires huge financial resources. For this purpose, either a part of the national income must be saved or the necessary funds be borrowed from other countries. Among these two means of capital formation, the role of domestic saving is so important that *Prof. Arthur Lewis* has defined the process of economic growth as the one that transforms a country from a 5% saver to a 15% saver. On the other hand, domestic savings though necessary are not sufficient for the purpose of capital formation, which involves three independent activities. In the first place, there should be a significant *increase in the volume of real savings* so that the resources that would have been used for consumption may be released for the purpose of capital formation. Secondly, there should be an efficient *finance and credit mechanism* so that the capital resources in the form of domestic savings or funds borrowed from abroad may be availed of by private investors or the government. Finally, there is the *act of investment itself*, so that the available resources are used for the production of capital goods.

Capital-Output Ratio

Apart from the ratio of capital formation to the aggregate national income, the growth of output also depends on the capital-output ratio. This ratio determines the rate at which the output of a country grows as a result of a given volume of the capital investment. A lower capital-output ratio leads to a comparatively higher rate of the growth of output as a result of a given volume of capital investment. It is difficult to estimate the capital-output ratio for a given economy. The reason is that the productivity of capital depends upon many factors like the degree of technological development associated with capital investment, efficiency of handling new types of equipment, quality of available managerial and organisational skills and pattern of capital investments. There is a broad agreement among economists that the capital-output ratio in under-developed countries is generally higher. According to the United Nations Experts Committee, the capital-output ratio for developing countries ranges from 2:1 to 5:1, implying that the capital is less productive in these countries. The reason for this reduced productivity is the relative inefficiency of their industries which produce capital goods. There is a great wastage of capital in the process of production due to the low level of technical knowledge in the under-developed countries.

The Rate of Population Growth

For effecting a significant improvement in the standard of living, the rate of capital formation and the consequent rate of the growth of output must be seen in relation to the rate of population growth. The rate of population growth is given by the difference between the birth rate and death rate of a country. In an underdeveloped economy, the population may be growing so fast that it may off-set even a relatively higher rate of capital formation and the resulting increase in the output. The population of a country can be both a helpful and a retarding factor in economic development. It depends on the size of population in relation to the natural and capital resources. If the size of the population is just enough to utilise these resources fully and efficiently, it contributes greatly to the development of the country concerned. On the other hand, if the size of the population is too large or too small in relation to natural and capital resources, it becomes an obstacle to economic growth. If the population of a country is too large, it contains a higher proportion of dependent population. A large dependent population implies a high expenditure on consumption and, therefore, low savings and investments. Since it is the responsibility of the government to provide the basic necessities of life to the citizens, a relatively higher population involves a higher expenditure to provide such necessities. As a result, less amount is available for investment in development projects. The data on the growth of population in India during the period 1951 – 2001 are presented in Table 4.3.1.1. As seen from the last column of this Table, the rate of population growth in India, though declining now, is still alarmingly high.

Table 4.3.1.1   Growth of population in India during the period 1951 – 2001.

Year	Population (million)	Increase in preceding decade (million)	Average annual growth rate (%)
1951	361	42	1.16
1961	439	78	1.76
1971	548	109	1.99
1981	685	137	2.00
1991	850	165	1.94
2001	1,025	175	1.71

According to the recommendations of the *Science Advisory Committee* of the Government of India, the most effective method of population control is the spread of education. To a very large extent, the qualitative aspect of a country's population is determined by the level of literacy. The higher the rate of literacy, the higher is the contribution of population to economic and social development. In India, the literacy rate among females is much lower than that among males, and females' education is most essential for the control of population growth and the development of the nation. The Government of India has taken various measures to control the birth rate under the family welfare programmes.

4.3.2 Non-Economic Parameters

The non-economic parameters provide as much motivation for economic development as the economic parameters. In fact, the difference in the growth rates of the developed and developing countries can be explained mainly on the basis of non-economic parameters like political sovereignty of the country, the type of government, the quality of administration and the political ideology of the government. The Government of India has put its faith in democratic planning for development. Equally important for the process of development are *social and cultural factors*. Each society has certain social institutions which have a strong bearing on economic growth. In India, the social institutions of caste, joint family system and the non-materialistic attitude of the people have proved to be some of the serious impediments to rapid development of the country. Any attempt at accelerating the development in India must aim at removing or at least weakening these age-old social institutions and bringing in about a fundamental change in the attitudes of the people. Likewise, rampant illiteracy among the people in most of the developing countries is another example of social and cultural factors which hinder economic development. A more detailed discussion on some of the important *non-economic parameters* for development is given below:

The Quality of Human Resources

In the past, economists used to talk of labour as if it was a well-understood uniform input into the process of production. Now it has been realized, that the typical adult worker is not just a basic factor of production, but a mixture of labour and a great deal of the so-called *embodied capital*. Thus, the output per head can be increased by increasing the quantity of the capital invested in a typical worker and improving the quality of labour.

All serious studies in industrial productivity point to the fact that advanced technical training pays not only the individual worker (or manager) who gets the training but also the society as a whole since it increases the per capita output. In addition to increased output, there are other general advantages derived by the society from an educated population. Various studies have shown that productivity improves with literacy and, in general, the longer the period of education a person has, the more adaptable he is to new challenges in this fast-changing world.

The Quantity of Labour

It is obvious that, for any given state of technical knowledge and the supply of other factors of production like machinery and raw materials, the size of the population can affect the level of per capita output. Every child born represents both a mouth to feed and a pair of hands to work. It is perfectly possible to speak of over-populated or under-populated economies, depending on whether the contribution of additional people to production would lower or raise the level of per capita income of the nation concerned.

Fig. 4.3.2.1 shows schematically the relationship between the per capita income and the population of a country and illustrates a case in which there is an *optimal population* for which the per capita income and the resulting standard of living are highest. After a certain optimal size, the pressure of more and more population applying given techniques to a given quantity of natural and capital resources brings into operation the law of diminishing returns. Once the output ceases to grow in proportion to the additional population, the per capita output must fall.

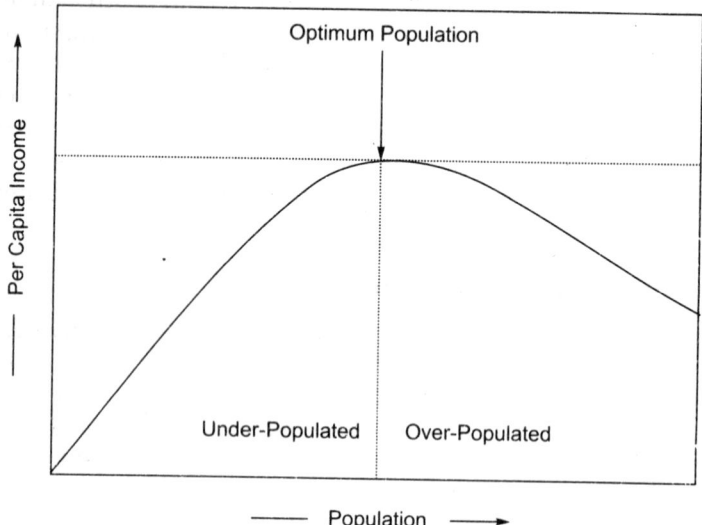

Fig. 4.3.2.1   The per capita income of a country as a function of its population

Many countries have had conscious policies which attempted to change the size of their populations. The US sought more immigrants in the nineteenth century to increase its population, as does Australia today. Similarly, Germany under the rule of Hitler offered incentives for having more children and Greece today is trying to stem the emigration of its people to Western European countries. These examples indicate a belief in insufficient population of the country concerned. In contrast, many of the developing countries in Latin America, Africa and Asia are trying to reduce and limit their population growth.

Invention and Innovation

New knowledge and inventions can significantly contribute to the economic development of a nation. In order to see this, let us assume that the proportion of the nation's resources devoted to the production of capital goods is just sufficient to replace the capital stock as and when it wears out. Now if the old capital goods were merely replaced in the same form, the capital stock would remain constant and there would be no increase in the national income. On the other hand, if there is a growth of knowledge, the old worn-out equipment would be replaced by different and more productive equipment. In this case, the national income will have to grow because of the growth of knowledge rather than the accumulation of more and more capital. This sort of increase in the national income can come about either by generating knowledge internally through research and development activities or by importing it from abroad through technical collaboration with multi-national companies. To some extent, developing countries can adapt the techniques that are already developed and used elsewhere. Developed countries, on the other hand, have the more difficult task of developing new techniques by research, invention and innovation. The historical importance of invention and innovation in contributing to economic growth is so well established that it hardly needs any elaboration. The assembly line and automation are radically transforming the means of production. Similarly, the invention of aeroplane has revolutionised transportation and electronic devices have come to dominate the communication industries. These innovations plus the less well-known but no less profound ones have created many new opportunities for investment and creation of wealth. The nature of goods and services consumed and the techniques employed to produce them are continually changing through research, development and innovation.

Social and Legal Institutions

Social habits of a country have a significant effect on its economic growth. Certain religious patterns are said to be more conducive to economic growth.

The well-known sociologist *Max Weber* argued that the Protestant ethic encouraged the acquisition of wealth and was thus more likely to encourage economic growth than an ethic that directed human activity away from the economic sphere. Although this hypothesis is not yet well-established, economists are interested in such relationships. On the other hand, if certain religious patterns or social habits tend to make economic growth more difficult, one cannot conclude that the religious and social structures should be changed to maximize the possibility of economic growth. In addition to religious and social institutions, the legal institutions of a country may also affect its economic growth. The pattern of ownership of land and natural resources naturally affect the way such resources are used which, in turn, affect their productivity. If agricultural land is divided into very small plots with each rural family having one such plot, it would be difficult to achieve the advantages of modern agricultural techniques. On the other hand, the concentration of land ownership in the hands of a few absentee land-owners, who are not much concerned to maximise their profits, may also be detrimental to economic growth. If the land lord's holdings are so large that he can obtain all the income he desires without using his land effectively, he may have little incentive to introduce advanced techniques. In many countries where this system (i.e., absentee land lords owning land) exists, land reform (which usually implies the confiscation, nationalization or communisation of land) becomes a necessary condition for growth.

4.3.3 The Role of International Trade

International trade plays a very important role in economic development since it allows a country to escape from its own limitations of natural and human resources and concentrates its efforts in the areas in which it has a genuine advantage. If there were no international trades, a developing country would have to grow on all fronts simultaneously. In that case, its growth could be seriously impaired by the limitations of natural resources and acquired human skills in many areas. A country bent on growing through an industrialisation policy may benefit if it can concentrate at first on light manufacturing and exporting consumer goods in return for capital goods made by heavy industries in more developed countries. In this way, a developing country can gain many of the benefits of more efficient production that it could not hope to match for a long time to come.

Among the other advantages of specialisations that international trade makes possible are the opportunities to take advantage of the economy of scale by producing far more goods than would be required to meet the domestic demand in a state of self-sufficiency. A further advantage that may be significant for a developing country is often called the advantage of "learning by doing". On the other hand, economic growth with a heavy dependence on the foreign trade often brings in the serious problem of the balance of payments in a world of fixed exchange rates. In a developing country, capital goods are often one of the main limitations to growth. In a *closed economy*, the problem of scarce capital appears as a resource problem since there are not enough resources to produce capital goods at a rate as fast as desired. In an *open economy* the same problem appears as a foreign exchange problem since there is not enough foreign exchange to buy all the imported capital goods that are desired for faster economic growth. In both cases, the problem is the same, i.e., it is very difficult to obtain a desired level of capital goods. One way is to make the capital goods at home; the other way is to make consumer goods at home and then sell them abroad in exchange for capital goods. A second problem in a developing country is related to the import of consumer goods, if the country's economy is an open one. As the country's productivity rises, disposable income and the standard of living also rise. In many developing countries, the goods produced at home are mainly in the necessity class with low margins of profit, whereas imported goods tend to have a higher profit margin. In such a situation, the rise in income that accompanies economic growth brings with it a shift in the pattern of consumer demand, with a larger proportion of consumers opting for the purchase of imported goods. Unless something happens to offset this shift in consumer demand, economic growth can be accompanied by an increasingly severe problem of the balance of payments. This problem can be offset in two ways. The first way is for the country to develop export commodities with rapidly expanding demand for them in foreign countries. In this case, the exports can expand rapidly to match the increasing imports. A second way is for its domestic growth to take place partly in the so-called "import-substitute" industries. Growth of industries that compete with imports can keep the rapidly expanding demand for luxury goods from being translated into an equally rapidly expanding demand for their imports.

4.4 INTERRELATIONSHIP BETWEEN SOCIAL, ECONOMIC AND SCIENTIFIC FACTORS

4.4.1 Science, Technology and Economy

Economic development is hardly possible without social change, and science, engineering and technology are the most important factors for changing a traditional society into a modern and developed one. Engineering is the applied science and technology is the applied engineering. This clearly shows that social, economic and scientific factors for development are highly interrelated. Technology, arising from scientific research followed by technological development, has been a prime mover in creating the kind of world in which we live today. From the shaping of the first stone tools, the discovery of the wheel, the lever and the plough and learning the use of fire, man has assiduously shaped science to serve his material needs. Science, therefore, is not a new phenomenon. What is different today is that the discovery of natural laws through scientific research has given a new dimension to technology. As a result, technology now has such a massive impact on our lives that it offers on the one hand an almost infinite promise to relieve poverty and provide healthy conditions of life, but on the other hand it also threatens our pattern of life, the global ecology and even the survival of the human race.

It is only during the last few decades that economists have paid serious attention to science, engineering and technology. Till then they implicitly assumed that new processes and products developed spontaneously in response to the interaction of economic forces, especially those of the market. It was in the early 1960s that the role of scientific research and development in economic growth became the subject of serious study by the economists. It was shown that in case of economic growth of the US during the first half of the twentieth century, only about 40% could be explained by increase in the traditional factors of production like capital and labour. The remaining 60% could be due to improvements in science and technology, materials, management skills and higher levels of education. This recognition of the importance of scientific and technological improvements and of the human and social factors for economic growth coincided with the enormous increase in research and development expenditures in all the industrialised countries after the Second World War. During the next two decades, it was generally assumed that science was a good thing in itself and that the more the scientific work done in a country, the greater would be its benefits to the economy and the society. Unfortunately, this sweeping generalisation proved false. Many examples can be cited of huge research and development efforts which proved technically successful, but economically disastrous. India, for example, has the third largest number of scientific and technical manpower in the world (after the US and Russia) and a very sophisticated scientific infrastructure, but we are still among the poorest countries in the world. The reason is that the process of technological innovations is extremely complex, and it is necessary that this complexity is thoroughly understood by political leaders, planners, industrial managers and scientists. Successful innovation demands initial technological novelty, either generated domestically or imported, as a first step. But careful selection from the large number of processes and products available in the world is an important prerequisite if the innovation has to serve the needs of a developing society. Some of the other factors which determine success or failure of a new enterprise are the availability of capital, a wise fiscal system (which protects the enterprise initially, but not to the extent of discouraging efficiency and high quality), managerial skills, successful marketing, general level of education, technical skills of the middle-level management, a good social climate and cultural traditions. Thus, it is evident that the social, economic and scientific factors for development are highly interrelated.

4.4.2 Development and Diversity of Nations

The interrelationship between social, economic and scientific factors becomes more evident when we consider the vast diversity among different nations and their equally diverse developmental needs. In the decades which followed the Second World War, it became obvious that the disparities between the rich and the poor nations had become intolerable. Deliberate attempts were made, therefore, to stimulate development in the so called Third World countries. During this period, it was recognised that technology was a key factor for development, whether it was production of food, industrialisation, expansion of infrastructure, conquest of tropical diseases, or removal of illiteracy. It seems obvious that the developing countries have to take advantage of the vast accumulation of technology in the developed countries. In this way, the developing countries would not have to repeat the laborious and costly process of research and development. At the same time, they could also avoid many of the ills of undirected technology which had appeared in the more industrialised countries in the form of environmental degradation, loss of work satisfaction and the social problems resulting from rapid urbanisation. Transfer of technology was seen as an obvious and major tool of development in the 1950s and also, in early 1960s. Today, only about 2% of the research and development for new technology take place in the Third World countries and this suggests that the developing countries must rely on the import of technology. Successful transfer of technology is a complex *social-economic process* with many variables and social conditions beyond the scope of mere introduction of packaged processes and know-how. Both the donors and the receivers of technology must assume that technology developed for a particular industrial, social and cultural environment in a developed country can be transferred smoothly and beneficially to a quite different environment in developing countries. The world's massive scientific and technological capacity is concentrated in a small number of highly industrialised countries, which are responsible for about 95% of the total research and development effort. Thus, there is a need to create a *new scientific and technological order* within the new international economic order if research and development activities are to become a driving force of economic development on a global scale.

4.5 ROLE OF SCIENCE AND TECHNOLOGY IN DEVELOPMENT

4.5.1 Science and Technology

In order to appreciate properly the rationale and relevance of adopting science and technology for development, it is necessary to draw a distinction between the two terms *"science"* and *"technology"*. As explained earlier, science and technology are closely related and highly interdependent. Policy for them, however, has to be distinct. Science is the result of man's restless quest to comprehend the phenomena of nature. By its very nature, development of science requires a long-term planning. No schedule of time can be fixed in advance for achieving a scientific breakthrough. Technology, on the other hand, is product or process specific and not universal. Unlike science, technology is not widely publicised and generally, not open to outsiders. The inventor guards the secrets of his technological breakthrough by getting a patent on it and thus preventing others from using the process developed by him. The possession of the patent on a technological development gives monopoly rights to the patent holder to derive commercial benefits from it for a fixed period. Technology is an essential input in all decisions relating to production in all sectors of the economy. As a result, technology is amenable to time-bound programmes, policies, strategies and planning on a continuing basis. All plans contain some specific projects with some technological contents in them. Projects launched for national development have to be supported by appropriate technological inputs that would improve productivity of both men and materials. In the process of modernisation, a country has to assimilate its own indigenous technology as well as the relevant imported technology. India has been doing so ever since it launched its ambitious *Five-Year Plans*. The role which relevant technology can play in improving productivity can be clearly seen from the experience of Punjab. A subsistence agriculture in Punjab was transformed into a commercial one, particularly after the Green Revolution in the 1960s, through the application of technology.

4.5.2 Technology and Development

Most technological innovations resulted from invention, and inventions were exploited when the market was ripe for them. Many ingenious but premature inventions were either ignored or had to wait until the market forces were favourable to them. With the discovery of electricity and advances in organic chemistry, great new possibilities were opened up for practical applications and these came from inventors and not from the deliberate efforts of scientific research. As science advanced, invention and innovation became increasingly sophisticated and, in the last fifty years several new industries could be established which were essentially based on scientific research applied through technological development. In the late 1960s, many academic research groups conducted intensive studies on the so-called "technological gap" between the US and the European countries. It was shown that no direct correlation existed between the proportion of the national expenditure of a country on science and technology and the rate of its economic growth. In fact, the extent of the national effort for advancement of technology seemed to have little effect on either economic or trade performance. It appeared that the diffusion of technology across national frontiers was sufficiently rapid to compensate for any inadequacies in the domestic research and development effort. This, however, appears to be true only for the *developed* countries above a certain level of technological threshold, with a mature R & D and industrial infrastructure. The above considerations do not apply to the diffusion or transfer of technology from the advanced countries to the less developed ones.

At the time of the first conference of the United Nations on *Science and Technology for the Benefit of Less Developed Areas*, held in Geneva in 1963, it was hoped that the economic gap between the industrialised and the developing countries could be rapidly narrowed by a massive transfer of technology from the former to the latter. This, however, proved marginally true. Massive transfer of technology has not yet fully eliminated hunger, disease and poverty in the developing countries of Africa, Asia and Latin America. Much of the new wealth created as a result of technology transfer and economic aid found its way to the already well-to-do people in many of these countries as a consequence of inequalities in their social systems. As a result, the great masses in the developing countries have not much benefited from the new technology, improved agriculture, and the inflow of capital aid and loans.

4.5.3 Obstacles to the Transfer of Technology

If the experience of the industrialised countries has been that science and technology have been major instruments for their economic development, one would assume that the same should be true for developing countries too. Certainly, great advances have been made in developing countries by the direct transfer of technology from advanced countries. For example, communications now form a world-wide network; certain devastating diseases like malaria, smallpox and tuberculosis have been controlled; and agriculture in the third world countries has made rapid progress by the use of fertilisers and high-yielding varieties of seeds.

The relative failure of the process of technology transfer is due to many causes, in addition to the lack of indigenous scientific and technological capacity. The profit motive of the donors of technology does not always harmonise with the basic needs of the receivers. Other causes of the failure of technology transfer lie in the social and political factors in the recipient nations. Another difficulty is the scarcity of capital. Also, some technologies are energy-intensive. In most Third World countries, unemployment and underemployment are widespread and, therefore, capital-intensive technologies are not suitable to them. Another obstacle to successful transfer of technology is the insufficient local availability of the necessary technical and managerial skills.

At present, the main political debate with regard to the use of science and technology for development is concentrated on the transfer process itself, and on the improvement in the access of advanced technology to the developing countries. The arguments used in the debate are highly politicised and centre around the effectiveness and motivation of the multinational corporations (MNCs) as the main agents of technology transfer. The MNCs are needed by the developing countries, but they are not encouraged. It is necessary that the multinationals and the developing countries come to terms. The multinational corporations will have to be encouraged to establish lasting relationship with the host country. The multinationals, as a matter of self-interest, will have to learn to balance the need for next year's profits with their long-term survival in the host country.

4.5.4 Brain Drain

Brain drain is the problem of skilled human resources flowing out of the developing countries, where they can make the greatest contribution to human welfare. Brain drain is a loss of vital human resource, and it is a loss to the developing country concerned without much compensation. Human capital is indispensable to a country's economic development. Moreover, the top-grade professional manpower is expensive to produce and it usually requires substantial public investment. Its loss through migration, therefore, represents a forced "gift" from a poor country which cannot afford it to a rich country. Such uncontrolled migration of brain power today favours the most advanced and affluent countries. Migration is an index of the extraordinary "pull" from abroad or the "push" from home or both. With respect to the developing nations, the brain drain is both the cause and the effect of all the economic, social and political factors which are grouped together under the widely-used term *"under-development"*. It is strange that the vitally needed manpower emigrates from a country in spite of its crying need for such human capital. The basic problem behind the phenomenon of brain drain is the lack of opportunity and absorptive capacity of the developing countries. There is an urgent need to formulate a global policy to control brain drain at both the receiving and sending ends, with the twin objectives of optimum resource allocation, and in the interest of the economic growth of the less developed countries of the world.

In order to solve effectively the problem of brain drain, several important measures must be taken. These include :

(1) Revision of salary structures.
(2) Providing increased professional opportunities.
(3) Rationalising manpower policies.
(4) Restructuring investment in education.
(5) Promotion of economic integration.
(6) Elimination of discrimination.

4.5.5 Science and Technology for Rural Development

Application of science and technology for rural development has acquired a special significance in India's development programmes. Rural people occupy the pivotal position in the context of anti-poverty programmes. It has been well recognised that the physical resources alone are not the only constraint for development. Even if physical resources are available, their inefficient utilisation can result in under-development. Thus, science and technology assume a greater significance in the context of rural development since they can increase the efficiency of resource utilisation. The areas of economy which can be substantially influenced through the adoption of science and technology can be grouped under production and development. Production efficiency in respect of the rural people relates to the following two aspects:

(1) Increasing efficiency of resource use for productive enterprise taken up by the rural people, either individually or in cooperative groups.
(2) Increasing efficiency of labour that the rural people provide to the economy through skill development.

Developments in science and technology can be fruitfully utilised for rural development through an improvement in the production efficiency of investment resources as well as that of the rural labour force, conservation of resources used by them and improvement in their health and living conditions. A great deal of effort has gone into the application of science and technology for rural development, but it has not yet produced desired results. While some technologies appropriate for rural development have been developed, their combined *package* (innovation, adoption, utilisation, human resources, raw materials and marketing) has not made substantial change in the productivity, income and living condition of the rural people in general.

The objectives should be :

(1) Development of technologies appropriate to rural areas in the existing socio-economic context of both producers and their customers.

(2) A "delivery system" or methodology to ensure the self-sustaining adoption and utilisation of these technologies.

(3) Maximisation of their benefits to the rural poor.

4.6 PLANNING—ITS OBJECTIVES AND ASSESSMENT

4.6.1 Characteristics of Planning

The essentials of planning are very well indicated in the definition of planning given by *Prof. Dickinson.* According to him, "Economic planning is the making of major economic decisions – what and how much is to be produced; how, when and where it is to be produced and to whom it is to be allocated – by the conscious decision of a comprehensive survey of the system as a whole". *Prof. Loucks* defines planning as "the shaping of all economic activities into group-defined spheres of action, which are rationally mapped out and fitted, as parts of a mosaic, into a coordinated whole, for the purpose of achieving certain nationally conceived and socially comprehensive goals". There are three important characteristics of planning. In the first place, there is a definite planning authority, whether it be the government or another body specially constituted like the Planning Commission in India. This planning authority is entrusted with the task of surveying the resource of the country, fixing up targets and laying down the methods for reaching these targets. If the government itself is the planning body, it also executes the plan thus formulated. It is not necessary that the planning authority should have supreme or dictatorial powers. It may be dictatorial, or it may work subject to the democratic control of a parliament as in India. It may even delegate some of its powers to other agencies, but the plans of all such agencies taken together must fit into the common pattern laid down by the planning authority. The second characteristic of planning is that the decisions are based on a survey of the economy as a whole. The planning authority acquires comprehensive knowledge of all resources that the economy possesses. Moreover, it keeps in view the needs and requirements of the economy as a whole, and not merely of some particular sectors of it, when it allocates the resources for various uses. In third place, the planning authority deliberately takes decisions with regard to the use to which various resources in the economy are put in. The economic life of the country concerned is not left to mere chance, or to the working of free competition. Planning implies the conscious and deliberate choice of economic priorities by the planning authority. An *unplanned economy* is characterised by the *absence* of these three characteristics. In an unplanned society, each individual is free, within such legal limits as may be laid down by the state, to make use of the resources at his command in the manner he thinks best. Naturally, he does not take a comprehensive view of the economy as a whole. He has, at his command, only a part of the total resources of the community. He uses these resources in such a way as to make the largest profit for himself, without any thought of the effects of his actions on others or on the economy of the country as a whole.

4.6.2 Shortcomings of Unplanned Economy

An unplanned economy according to some economists, is not really "planless"; in fact, it tends to be even more "planned". This is so because of the working of the price system under the conditions of free competition. What the society demands or not, at a particular time, is reflected in the price of different commodities. The price of goods and services for which demand increases tends to go up, and the price for which the demand decreases tends to go down. In search for maximum profits, there is a natural tendency to make the best allocation of the available resources. However, the working of an unplanned economy is not really as smooth as the supporters of free enterprise make it out to be. Unplanned economy has the following shortcomings:

Wastefulness: It is not entirely correct to say that an unplanned or "free" economy tends to make the best use of available economic resources. The guiding force in the free enterprise system is profit and not the maximisation of welfare of the society as a whole. It is assumed by the supporters of free enterprise that the search for private profits automatically leads to the optimum utilisation of resources for the community. But this is not necessarily true in general. In an unplanned economy, the character of production of the goods and services is not determined by the *needs* of the people, but by *demand,* i.e., by what people are prepared to buy at certain prices. This often leads to the production of luxuries for the rich rather than necessities for the poor. Moreover, consumers' demands often express not what they really need, but what they *think* they need. By the clever use of advertising, the consumers are persuaded to buy various things, whether they really need them or not. Thus, the resources of the society may be spent on producing luxury items in preference to the necessary ones. Another characteristic waste of resources in a free economy arises due to competitive advertising, which results in duplication of production of the same thing, with minor differences (e.g., many different brands of soft drinks, ice-cream or mineral water). Resources are further wasted in persuading the customers to buy one brand rather than another. The greatest waste in the unplanned economy occurs due to the fact that the resources cannot be utilised in a coordinated manner since no single enterprise has any comprehensive knowledge of the total resources of the nation.

Uneven Distribution of Wealth: The system of free enterprise generally leads to concentration of wealth in the hands of a few people. The reason is that free enterprise system is based on the principle of private property and the right of inheritance. It is, therefore, the owners of the means of production who not only decide what should be produced, but how to make and get the profits to increase their wealth further. This results in the tendency for the rich people to get richer. Theoretically, any person who has the requisite business ability and initiative can start his own enterprise and receive his share of wealth. In actual practice, the opportunities for doing so are severely limited by various factors. The net result is inequality in the distribution of wealth and, in most cases, it is a gross inequality.

Economic Instability: The unplanned system tends to be economically unstable. One of the most characteristic features of the capitalistic system of economy has been the business cycle. Recurring economic booms and depressions or recessions cause dislocation of economic life, acute misery and suffering. When there are too large disturbances in the economy, people suffer as a result of scarcity and high prices in the inflationary or boom periods, and unemployment and economic ruin during depressions and recessions. The necessity of maintaining stability in the levels of business and employment points to the importance of economic planning and centralised monetary control.

Monopoly: Unplanned economies are usually characterised by the growth of monopolies. The virtues claimed for the system of free enterprise are based on the assumption of fair competition. In fact, monopolies develop to such a large extent in the free enterprise economies that assumption of fair competition proves unrealistic. As a result, the best utilisation of resources, supposed to be a major plus point for the free enterprise system, cannot really take place.

Economic Backwardness: Many countries which are now underdeveloped, are seeking to achieve rapid economic growth. These countries cannot do so if they merely depend on the automatic working of economic forces-the very forces that have kept them backward so far. For these underdeveloped countries, it is essential to plan optimum utilisation of their resources for achievement of their development objectives. Economic planning is, therefore, essential and vital for the developing countries trying to achieve rapid economic growth. A belief in the importance and efficacy of economic planning has grown rapidly among economists, social scientists and political leaders. It is generally believed that central control and planning lead to a rational use of economic resources of a country. In addition, central planning is likely to bring greater stability in the economy and, therefore, prevent sharp fluctuations in the level of economic activity and employment.

4.6.3 Forms of Planning

The adoption of planning by the former USSR after the Russian Revolution in 1917 gave impression that all planning must be socialistic. Socialism certainly involves planning, but all plannings need not follow the socialistic pattern. Planning is possible to some extent even in a capitalistic country, but the methods adopted for planning in such countries are quite different. The characteristic feature of a planned economy in a *socialistic* country is that all resources of the nation and all instruments of production are owned by the state and are utilised by it to achieve the goals set for economic development of the nation. Private enterprise has little or no place in a socialistic economy.

A *capitalistic economy* may also be planned. The planning in such economies may be on a very elementary level as in the case of the US, or it may be very elaborate as in the case of Nazi Germany under Hitler. The extent of planning in the capitalist US has been very small. A small step towards planning was taken in the US in 1946 by the passage of the Employment Act with objective of "creating and maintaining in a manner calculated to foster and promote free competition, enterprise and general welfare, under which there will be useful employment opportunities including self-employment for the able, willing, and seeking to work, and to promote maximum employment, production and purchasing power". Although the Employment Act has led to some integration of measures relating to employment, it is not a "planning" in the full sense of the term. The American economy is basically an unplanned economy despite some measures relating to the Government intervention in economic matters. In the then Nazi Germany and Fascist Italy, there was a growth of *planned capitalism*. The ownership of the means of production was left in private hands, but all production, pricing and trade were carried on according to the directions given by the Government. The entire economic life was planned and directed by the government to achieve certain objectives.

Between the two extremes of completely planned socialist economy and almost unplanned capitalist economy, there lies a spectrum of *mixed economy* with varying degrees of planning. The planning in a mixed economy envisages coexistence of both the public sector and the private sector. In other words, while certain basic aspects of economic life are directly under the ownership and management of the estate, other less important aspects may be left to private enterprise. Such a planning has been adopted by India after independence. Somewhat similar methods of planning were adopted in Great Britain by the Labour Government after the Second World War.

4.6.4 Methods of Planning

The aim of the planning authority is to ensure that the resources of the country are utilised in such a manner so as to achieve certain predetermined objectives. These objectives can be achieved by a number of different methods:

(1) Social ownership of all or some of the means of production.
(2) Issue of orders and directives as to how people will use the resources at their disposal.
(3) Enforcement of various controls relating to economic matters.
(4) Use of fiscal, monetary and price policy.

A *socialist state* itself owns and operates all the resources of the economy. It can use them in such a manner as to achieve the goals it sets out in the national development plan and direct the resources into various pre-determined channels. A *capitalist state* may have to adopt any one or a combination of the methods. A capitalist state may try to mould the working of a part or the whole of the economy by issuing orders or directives with regard to its functioning, and by imposing controls on the production, movement, sale, consumption and price of various goods. Alternately, it may try to create conditions in which people find it in their own interest to act in such a way that the objectives of the plan can be achieved. The government may use fiscal policy (taxation, borrowing, expenditure, etc.) or monetary policy (determining the volume and rate of credit) to encourage certain economic activities and discourage some others. The government may also influence the supply and / or demand of various commodities, and thus increase or decrease their production, consumption and price. The first of these methods (issue of orders, imposition of controls, etc.) is known as "*planning by direction*" while the second method (use of fiscal and monetary policy) is called "*planning by inducement*". In a mixed economy, the most important sectors of the economy are owned by the government itself and can be operated in accordance with the development plan. There is also a private sector in a mixed economy, for which the government may use either the direction or the inducement method. The major implications of these two methods of planning are considered below:

Planning by Direction: This implies that the planning authority, having decided upon the targets and objectives to be achieved, proceeds to direct or order the people to act in certain ways, or to desist from acting in certain other ways, so that those targets may be reached. The planning authority may impose restrictions on the movement, imports or exports of certain commodities. It may impose controls on the price of certain commodities. It may impose rationing of goods in short supply. It may order the banks to lend money on easy terms for certain purposes and it may compel the producers to produce certain things, or may prohibit them from producing others. In planning by direction, the planning authority allocates the economic resources (raw materials, labour, capital goods, etc.) to different industries and occupations, and it allocates the total product of economy (consumption, export or capital formation). In short, the planning authority issues the directives that such and such things shall or shall not be done. Planning by direction has some drawbacks. In the first place, it requires very accurate determination of how exactly each unit of various resources shall be used. Such an accurate and complete calculation may be very difficult. Secondly, even if the required accurate calculation can be made, a change in conditions may upset them. Prolonged labour strikes, crop failures, reduction in imports due to wars, etc. may hold up not merely one part of the plan but the whole of it. A third difficulty in planning by direction arises from the fact that all parts of the plan are more or less interlinked after elaborate calculations and estimates. As a result, any change in one part of the plan leads to a chain of alterations and modifications in many other parts of it. Due to this, the planning authority tends to resist any attempts at revising or modifying the plan. This planning by direction cannot be flexible. A fourth drawback of this method of planning is that a huge bureaucratic machinery is needed to formulate and implement the plan. Such a machinery is likely to be expensive. Finally, planning by direction tends to be authoritarian, imposing restrictions on the people and curtailing their freedom of choice as producers, workers and customers.

Planning by Inducement: The other method of planning is through inducement. In this case, the planning authority induces the people to act in certain desired ways through fiscal and monetary policies and through the price mechanism. Fiscal and monetary measures can obviously play an important role in planning since the extent and nature of production can be influenced through taxation and expenditure. When it is desired to increase the production of a commodity, the government may give subsidy for its production, grant it a protection, or give it some tax concessions. Monetary measures may be taken to encourage or discourage the productive activity in general and the desired objectives may be achieved more smoothly through use of the market and price mechanisms. A judicious use of the pricing system enables the producers to produce what is most wanted in the market in economical manner. The pricing system also gives each individual a general command over his fair share of the community's resources. Thus by manipulating and controlling the market, the state can do all the planning and through inducements can make the producers produce commodities that are socially desirable.

In spite of these advantages, planning by inducement has some drawbacks. In the first place, this method of planning, by its very nature, does not lead to an exact match between supply and demand. As a result, shortages and surpluses are always likely to remain. Secondly, the mobility of the factors of production is often so low that no amount of inducement by itself may be sufficient to bring about a change in a short time. For example, the prospects of high wage in far off places may not induce villagers for their relocation. In the third place, if certain essential resources are in short supply, no amount of inducement may make people reduce their consumption. Finally, too much dependence on monetary measures may lead to inflation or deflation.

Mixed Planning: In actual practice, neither planning by direction nor planning by inducement gives the best results. Both have their own advantages and disadvantages. In case of planning by direction, there is a limit beyond which the control measures (prices, movements of goods, rationing, etc.) cannot be worked out successfully. Similarly, in case of planning by inducement (through the market mechanism), there arises a situation where direct controls have to be imposed. It follows, therefore, that both methods of planning must be used simultaneously in such a way so that they can complement each other. For example, if there is a shortage of any commodity, measures may be taken to ration its consumption on the one hand, and to encourage its production on the other hand. In normal conditions, the budget should be the principal instrument of planning, and it will have to be supplemented by controls in all those parts of the economy when there is market disequilibrium between demand and supply.

4.6.5 Objectives of Planning

Economic planning has become a craze in modern times. It has been especially recommended for the under-developed countries to cure them of all economic ills. The fundamental objective of planning is to accelerate economic development of the country by bringing about an optimum utilization of its natural and human resources, so that the masses can have a reasonably high standard of living. The objectives of planning, however, are not the same for all countries, nor are they the same for any country at all times. What precisely are the objectives of planning depend on the stage of economic development of the country concerned, the socio-economic conditions prevalent there at the time of planning, and the requirements for a particular situation. Keeping the broad objective of accelerating economic growth and of raising the standard of living of the people, the major objectives may be listed as follows:

(1) Achieving full employment.
(2) Maximising national income and raising living standards.
(3) Rapid industrialization.

(4) Achieving self-sufficiency in food and raw materials.

(5) Reducing inequalities.

(6) Redressing imbalances in the economy.

It may be noted, however, that these objectives are interrelated and complementary, and not mutually exclusive of one another.

(1) Achieving Full Employment: One of the major objectives of planning especially in economically advanced countries, is to provide full employment to its citizens. Unemployment is one of the by-products of capitalism, and is considered by many to be the biggest curse of a modern capitalist society. In order to end unemployment, the development plans are directed to those directions and those sectors of the economy where there is more unemployment. Towards this end, the state can redistribute the labour force and productive resources, and thus create more jobs in those sectors. Planning in a developing country like India may not be able to create the conditions for full employment in the near future, but it can reduce the incidence of unemployment.

(2) Maximising National Income: Another laudable objective of economic planning is to maximize the national income and to raise the living standard of the people. The planners pursue this objective relentlessly. The reason is that only planned efforts can raise the national income appreciably. The unplanned development efforts do not bring any substantial progress in this direction since the productive efforts are frittered away in an uncoordinated economic activity.

(3) Rapid Industrialisation: This objective assumes more importance in the case of those countries which have been left behind in the race for industrialisation. It has been realized that industrialization plays a very important role in raising the national income and in solving the problem of unemployment. It is difficult for a country to achieve prosperity by merely confining itself to agriculture. The economies which are predominantly agricultural are bound to remain backward unless attention is paid to industrialisation.

(4) Achieving Self-Sufficiency: As a first step towards more systematic and intensive planning, it is important to make the country self-sufficient in the matter of food and essential raw materials. This self-sufficiency provides a solid base for the country and prepares it for further development through industrialisation. If a country is not self-sufficient in the matter of food and raw materials, mere political freedom may prove to be a myth.

(5) Reducing Inequalities: It is generally realised that political equality is meaningless unless it is accompanied by economic equality. Glaring inequalities of wealth, income and opportunities are against the spirit of

democracy. Bridging the gulf between the rich and the poor is one of the most important objectives of planning.

(6) Redressing Imbalances in the Economy: Often it is found that the economic development in a country is lopsided. An economy, for example, may be predominantly agricultural, or it may depend too much on the export of oil or minerals. In our own country, nearly 75% of the people are dependent on agriculture and related activities. This is an example of an unbalanced economy. To lend stability to the economy, it becomes essential to reduce this imbalance. If there are such imbalances in the economy of a country, removing them becomes an important objective of planning. In addition, there is usually a regional imbalance in respect of industrialisation in many of the developing countries. When this is the case, the plans are aimed to correct such regional imbalances too.

Considering all these objectives of planning, it is evident that there is no need for the planning authority to adopt only one objective. That may be possible for the industrially advanced countries which do not have much to seek now in the way of economic development. The developing countries, on the other hand, usually suffer from many shortcomings, and their plans have to pay attention to many objectives simultaneously. As a result, the plans in developing countries are generally *multi-objective* in nature. It is necessary, however, to confine to a few major objectives at any one time, and leave other objectives to the next plan, lest the planning efforts may get diffused.

4.6.6 Planning in India and its Assessment

The first attempt at systematic planning in India was made by Bharat Ratna Sir M. Visvesvaraya, an eminent engineer and a great visionary, when he published in 1934 a book entitled *"Planned Economy for India"*. Three years later, the Indian National Congress set up the National Planning Committee under the Chairmanship of Pandit Jawaharlal Nehru, which submitted its report in 1948. In the meantime, eight leading industrialists of Bombay came out in 1943 with *"A Plan for Economic Development in India"*, which was popularly known as the *"Bombay Plan"*. This plan aimed at increasing the per capita income in India by 100% within a period of 15 years. This was sought to be achieved by raising the agricultural production by 100% and industrial production by 500%. The Bombay Plan accorded top priority to basic industries. Simultaneously with the Bombay Plan, Mr. M.N. Roy released a ten-year *People's Plan*, envisaging a total outlay of Rs.15,000 crore. In contrast to the Bombay Plan, the People's Plan gave highest priority to agriculture and consumer goods industries, thus relegating basic industries to a secondary place. Apart from these non-official attempts at planning, the Government of India also realized the need for planning and accordingly a

Department of Planning and Development was set up in 1944. This Department drew up both short-term and long-term plans, the former for the restoration of economic normalcy after the end of the Second World War and the latter for the country's economic reconstruction and development.

The real beginnings of planning in India were made only after the independence. The *Indian Planning Commision* was established in March, 1950 with Pandit Jawaharlal Nehru as its first Chairman. The Commission presented the draft outline of the *First Five-Year Plan* in July, 1951 to cover the period from April, 1951 to March, 1956. The First Five-Year Plan was duly followed by the Second, the Third and the subsequent Five Year Plans. The Third Five-Year Plan ended on March 31, 1966. Ordinarily, the Fourth Five-Year Plan should have commenced on April 1, 1966. The Indo-Pakistan War in 1965, devaluation of Rupee, wide-spread drought and the economic recession held up the finalisation of the Fourth Plan. It was only in August, 1966 that the Draft Outline of the Fourth Plan was released and that proved to be still-born. Due to uncertainty of foreign aid, the Draft Outline was abandoned, and there followed a three-year period of "deadly pause", which was termed as "Plan Holiday". During this period (1966-67 to 1968-69), only *Annual Plans* were formulated to carry on the work of planning.

Then followed the Fourth Plan (1969-70 to 1973-74) and the Fifth Plan, which was originally scheduled for completion at the end of 1978-79. But the Janata Government that came to power in 1977 terminated the Fifth Five-Year Plan one year earlier and drew up it own Draft Five-Year Plan (1978-79 to 1982-83), based on the concept of rolling plan and reflecting its own priorities. The Janata Plan was terminated after a period of just one year when the new Sixth Five-Year Plan (1980-81 to 1984-85) was formulated by the newly elected Congress Government.

4.6.7 Assessment of India's Five-Year Plans

Planning in India derives its objectives from the *Directive Principles of State Policy* enshrined in the *Constitution of India*. The *Planning Commision* was set up in 1950 to prepare the blueprint of development, taking an overall view of the needs and resources of the country. In the past, the economic planning in India envisaged a growing public sector with massive investments in the basic and heavy industries. Now there is less emphasis on the public sector. The current thinking in India on planning in general, is that it should be indicative in nature.

First Plan (1951-56): The First Plan, with a total outlay of Rs. 2,378 crore, was a rather haphazard affair since the Planning Commission did not have reliable statistics on the resources. Moreover, the plan had also to be correlated to the activities of various other departments and ministries of the Government. The result was a patchwork of various isolated projects. In spite of these shortcomings, the plan had a national character, and it was based on a rational hypothesis, viz., economic development through planned efforts. The First Plan laid emphasis on agriculture, irrigation, power and transport in order to provide the necessary infrastructures for rapid industrialisation. The plan was more successful than expected, mainly because it was supported by two good harvests.

Second Plan (1956-61): In sharp contrast to the First Plan, the Second Five-Year Plan was a big-leap forward. This plan laid special emphasis on heavy industries like steel. The *Industrial Policy Resolution* was amended so as to shift the primary responsibility for development on the public sector, while the private sector was left to handle the consumer goods industries. However, the huge quantities of import that the Second Plan envisaged in both public and private sectors, practically wiped out India's accumulated foreign exchange (nearly Rs.500 crore) in just two years and forced the country to seek extensive foreign aid. During the plan period, agriculture and small-scale industries remained sluggish.

Third Plan(1961-66): The Third Plan generated high expectations following an overall growth of Indian economy during the period of first two plans. The aim of the Third Five-Year Plan was to establish a self-sustaining and vibrant economy in the country. Since the internal resources had already been exploited to the utmost, the Third Plan had to rely heavily on the foreign aid. During the Third Plan, the national income at 1960-61 prices rose by 20% in the first four years, but it registered a decline of 5.6% in the last year of the plan due to wide-spread drought, Indo-Pakistan war in 1965 and other adverse circumstances. As a result, the per capita real income in 1965-66 was roughly the same as it was in 1960-61. A growing trade deficit and mounting debt obligations led to more borrowings from the International Monetary Fund (IMF). There was devaluation of Rupee in June,1966 and the plan did not serve the desired purpose.

Interim Planning (1966-69): Since the Third Five-Year Plan had gone awry, the process of planning itself became discredited in the eyes of many people. As a result, there were demands from many different quarters to declare a *Plan Holiday*. However, neither the Government nor the Planning Commission admitted the failure of the Third Plan. They refused to entertain the demand for a Plan Holiday and went ahead to prepare the Draft for the Fourth Plan, which was scheduled to start from 1966-67. However, the Indian economy had degenerated to such an extent that the Fourth Plan could not be started in 1966. Instead, the planning was done on an annual basis as a stop-gap arrangement, following the concept of Rolling Plans. The arrangement of Annual Plans continued from 1966 to 1969. A common man felt no change during these years.

Fourth Plan (1969-74):   The Fourth Five-Year Plan officially commenced on April 1, 1969. The main objective of the Fourth Plan was economic growth with stability. It was expected that agriculture would grow at a rate of 5% per annum, to lead the overall economic growth. The planners hoped that a higher rate of growth in agriculture would lead to a chain reaction of higher growth rate in other sectors of the economy . Thus, the target for the growth rate of industry was set at 9% per annum. Altogether, the national income was expected to rise at the rate of 5.5% per annum in the Fourth Plan. Allowing growth of population at the rate of approximately 2.5% per annum, the per capita income was expected to increase at the rate of about 3% per annum to show a total rise of about 16% during the period of the Fourth Plan. Unfortunately, the goal could not be achieved.

Fifth Plan (1974-79):   The Draft of the Fifth Plan, as originally drawn up, was part of a long-term Prospective Plan covering a period of ten years from 1974-75 to 1984-85. The Fifth Plan attempted to coordinate various sectors of the economy in terms of the new slogan *Garibi Hatao* ("Remove the Poverty"). The economy was expected to achieve a long-term rate of growth of 6.2% per annum on a self-sustaining basis. By the time the Fifth Plan was approved by the National Development Council in September, 1976, its premises had already become obsolete. As a result, the total outlay of the Fifth Plan had to be increased from Rs. 37,463 crore to Rs. 39,303 crore. The belated attempt had an inglorious end. When the Janata Government came to power at the centre, the Plan was scrapped. The Janata Government reconstituted the Planning Commission and announced a new strategy in planning. This strategy involved a change in the objective and in the pattern of planning. The objective of the Plan was laid down as *Growth for Social Justice* instead of *Growth with Social Justice.* Similarly, the new pattern of planning was laid down as the *Rolling Plan,* which implied that the performance of the Plan was to be assessed every year, and a new Plan based on this assessment would be prepared for the next year. The Rolling Plan was started with an Annual Plan for the year 1978-79 as a continuation of the terminated Fifth Five-Year Plan. The plan could help the poor people to some extent.

Sixth Plan (1980-85):   The Sixth Five-Year Plan was formulated after taking into account the achievements and shortcomings of the previous three decades of planning. For the Sixth plan, the actual expenditure stood at Rs.109,291.7 crore. The average annual growth rate for the Sixth Plan worked out to 5.2% and it could not be achieved.

Seventh Plan (1985-90):   The Seventh Five-Year Plan, which envisaged an aggregate outlay of Rs. 348,148 crore with a public sector outlay of Rs. 180,000 crore, ended with an average rate of growth of the GDP (Gross Domestic Product) at 5.3% per annum. This rate of growth was well above the targeted rate of 5% per annum. The final year (1989-90) of the Seventh

Plan saw the growth of the national income by 4%, which was largely contributed by the secondary (manufacturing) and tertiary (services) sectors of the economy.

Eighth Plan (1992-97): The Eighth Five-Year Plan recognised the need for a reorientation of planning, keeping in view the process of economic reforms and the restructuring of the economy. The Eighth Plan emphasised :

(1) Human resources development as the main focus of planning.
(2) A large economic space for the private sector.
(3) Development of physical and social infrastructures by the public sector,.allowing at the same time the private sector also to participate.
(4) A greater role for the market to infuse economic efficiency even in the working of the public sector.

The Eighth Plan envisaged an average growth rate of 5.6% per annum during the plan period. An outlay of Rs.798,000 crore was projected for the five-year period of the Plan. Out of this, the investment for the public sector was Rs. 361,000 crore (about 45% of the total outlay). Consistent with the expected resource position, the size of the Plans of the states and the union territories was projected at Rs. 186,325 crore. Again, the growth rate could not be achieved.

Ninth Plan (1997-2002): The objectives of the Ninth Five-Year Plan were evolved from the *Common Minimum Programme* of the Central Government and the Chief Ministers' Conference. The main objectives were as follows :

(1) Priority to agriculture and rural development with a view to generating productive employment and eradication of poverty.
(2) Accelerating rate of growth of the economy with stable prices.
(3) Ensuring food and nutritional security for the vulnerable section of the society.
(4) Providing basic minimum services of safe drinking water, primary health-care facilities, universal primary education, shelter and connectivity to all.
(5) Containing the growth rate of population.
(6) Ensuring environmental sustainability of the development at process through participation of the people.
(7) Empowerment of women and the socially disadvantaged groups (Scheduled Castes, Scheduled Tribes, Other Backward Classes and Minorities).
(8) Promoting and developing Panchayati Raj and Cooperatives.
(9) Strengthening efforts to build self-reliance.

According to rough calculations, the total outlay of the Ninth Plan was about Rs. 880,000 crore. Accordingly, the private investment in the Plan had to be around Rs. 550,000 crore as against the public sector's investment of Rs.330,000 crore. The Ninth Plan envisaged a growth rate of the GDP at 7% per annum. Although the growth rate could not be achieved, the standard of living of the people could be increased.

Tenth Plan (2002-07): The current economic trends in India show a dismal picture. In 2002-03, the economy grew by only 5.2% as against the official revised estimate of 6%. Apart from the negligible growth in agriculture, the decline continued in industry and it spread even to services. Moreover, the economic growth could not make visible gains in terms of poverty eradication or guarantee of providing the basic minimum needs to rural India. Keeping in view the disappointing situation, the Approach Paper for the Tenth Five-Year Plan, drawing lessons from the past, outlines a strategy which integrates higher growth rate with equity and social justice. The aim of the Tenth Plan is to set a higher growth rate for the economy at 8% per annum. The Planning Commission has set specific monitorable targets for the Tenth Plan such as reduction of poverty, more employment, universal access to primary education, increase in the literacy to 72%, lowering of the child mortality rate and provision of potable drinking water to all the villages.

MODEL QUESTIONS

(Essay/Long Type)

4.1 Explain in brief some of the major aspects of development.

4.2 List the impediments to development. Explain these in brief.

4.3 Define and explain an under-developed economy. What are the basic characteristics of under-developed countries?

4.4 List and explain various parameters for development.

4.5 "International trade plays an important role in economic development"–Justify the statement.

4.6 Write a brief note on interrelationship between social, economic and scientific factors.

4.7 Explain the role of science and technology in development.

4.8 Write a note on the obstacles to Transfer of Technology.

4.9 Explain the role of science and technology in rural development.

4.10 Define planning. What are its characteristics?

4.11 What are the shortcomings of unplanned economy? Explain in brief.

4.12 Explain the objectives of planning.

4.13 Give an assessment of India's Five-Year Plans.

(Objective/Short Type)

4.14 Define Science, Engineering and Technology.

4.15 Define the "rate of population growth".

4.16 What is the most effective method of population control ?

4.17 What is the range of capital-output ratio for the developing countries?

4.18 What is the meaning of "planning by direction" ?

4.19 Define "Plan Holiday".

4.20 What is a Rolling Plan ?

4.21 During which Five-Year Plan period, maximum emphasis was laid on steel industries ?

4.22 What is Green Revolution ?

4.23 When was the slogan "Garibi Hatao" given ?

4.24 Pre-machine technology includes :

    (a) Basket weaving   (b) Pottery making

    (c) Only (a)   (d) Both.

4.25 Functional differentiation is represented by :

    (a) Trade unions   (b) Political parties

    (c) Both   (d) Neither.

4.26 Match the following :

    (A) Bharat Ratna Sir   (a) Mixed economy
           M.Visvesvaraya

    (B) India   (b) High mass consumption

    (C) Germany   (c) Under-developed economy

    (D) Colin Clark   (d) Planned Economy for India

4.27 Write true or false :

The caste system allows vertical social mobility.

4.28 Write true or false :

Advertising is an important aspect of economic development.

4.29 Name the odd ones :

    (a) Economic instability

    (b) Multinational corporations

    (c) Uneven distribution of wealth

    (d) Economic backwardness.

4.30 Fill up the blanks :

Some of the impediments to development are :

lack of skills, _____, _____ and premature politicization.

5
Chapter

Technology Assessment

5.1 HISTORICAL DEVELOPMENT OF SCIENCE

5.1.1 Science and Society

The term *"science"* is generally applied to any discipline of knowledge or body of systematic principles, and especially to those disciplines like physical sciences whose principles are universally accepted. This definition of science, however, does not permit one to speak of science as a unity, but only as a generic name for a number of independent and highly diverse disciplines. Fortunately, there is another definition of science, based on a historical analysis of the development of various sciences in human culture. This definition regards science as an organic unity, expressing the same method in all its branches and manifesting its effects on social and cultural life as a single force. Science may also be defined as a far flung system of knowledge based on concepts which may allow it to serve as a theoretical basis for practical techniques. It was in practical physical techniques that man first came to appreciate the need of a representation of the universe of perception in terms of basic concepts and relations which could be relied upon to remain invariant with time. On the basis of this definition of science and analysis of the place of scientific thought and scientific approach in the wide spectrum of human thought, it is possible to trace intelligently the historical development of science and its interplay with other phases of culture. Science had a very late development in human civilisation. Religion, social organisation and even the fine arts have a continuity reaching back a hundred thousand years into the past, but the thread of science does not go back to more than four or five thousand years. The greatest part of history of ancient science pertains primarily to mathematical and astronomical sciences. As applied to natural and physical phenomena, science cannot be said to have become established before the seventeenth century and its era of intensive development dates only from the nineteenth century. The main reason for the late development of science is that primitive cultures reflected an antiscientific tendency and an inherently false approach, which had to be outgrown before science could begin and progress.

5.1.2 Science in Ancient Egypt and Babylon

Before a scientific mentality could emerge in a culture, there was required a long and slow transformation of its physical and institutional conditions. The progress of social organisation created a social class with sufficient leisure and curiosity to conduct a disinterested inquiry in certain directions. This state of things corresponds to the great civilisations of ancient Egypt and Babylon. In both of these civilisations, in the midst of a general magical mentality, there existed highly developed mathematical science, which went far beyond the practical needs. This mathematical science embraced not only arithmetic but also geometry. The history of the concept of numbers reveals its roots in concrete numeration, i.e., in telling of objects in a one-to-one correspondence against a set of standard objects. A similar development occurred, at a much later stage, with regard to the tracing of lines and figures which could be superimposed on concrete objects to determine their size and shape. These efforts gave birth to the science of geometry. The extent of development of numerical and geometrical analysis depended on conditions of leisure and interest in the subject. One is amazed to find that, as early as 2500 B.C., the Babylonia priests had advanced in the analysis of numerical operations to the point where they were able to solve simultaneous quadratic equations with two unknowns. Similarly, in case of geometrical science, they had atleast an implicit knowledge of the Pythagorean Theorem two thousand years before Pythagoral was born. These extraordinary mathematical developments in Babylon and in ancient Egypt failed to have any rational repercussion on the prevailing general mentality. In Babylon, mathematical developments combined with magical and superstitious ideas, gave birth to astrology. These were passed on to Greece after the entry of Alexander the Great into Babylon. The type of scientific developments represented by ancient Egypt and Babylon has been repeated several times in India, China, Japan and among the Mayas of North America. It serves as an evidence to show that the potentialities of science are not confined to any one race but may be regarded as belonging to the whole of human race. In face, however, only one civilisation has been able to produce a fully developed scientific tradition, and that is the so-called *western civilisation*, which traces its origin to the Greek civilisation.

5.1.3 Science in Greek Civilisation

The historic importance of the Greeks from the point of view of science lies in two directions. In the first place, they laid down the programme and formal outline of the whole range of knowledge, including not only the positive sciences, but also the philosophical disciplines. Secondly, they built up and developed the science of geometry and started applying it to the motion of heavenly bodies and to preliminary problems of mechanics. Greek philosophical thought contributed an important element into modern science. The dawn of European science has traditionally been located among the Greek philosophers in the sixth and fifth centuries B.C. Their work is known only through references and brief quotations made by authors who came much later. The two learned arts which gained momentum in Greece by the later part of the 5th century B.C. were medicine and geometry. The practice of medicine was atleast attempting to apply disciplined method in observation and inference, while geometry was accumulating a body of results about relations among particular figures, and was approaching the problem of logical structures and definitions. Plato, who lived in the early 4th century B.C., was the earliest Greek philosopher. He was a powerful propagandist for mathematics. In his famous book *Republic*, Plato argues that geometry prepares the mind for the discourse about the real ideas and thence, to wisdom and illumination. In another book entitled *Timaeus*, he gives a theory of music in terms of simple ratios, and a survey of the theories of physics and physiology accepted during that period. Aristotle, who also lived in the 4th century B.C., was one of the world's first and the greatest scholar. His interests ranged over the entire natural and human world including ethics and metaphysics. Through accurate observations, he created biological science and taxonomy much like those in use now. Aristotle also organised cooperative research for large-scale studies. He was a master of the scholarly method. In each study, he would define the area and its major problems, critically review the work of his predecessors, and then proceed by experience and reason to develop his arguments. Aristotle started his career as a disciple of Plato, but eventually came to disagree with Plato on the fundamentals. Aristotle's biological studies culminated in the problems of generation. He tried to explain the working of the first cause of all physical phenomena through the study of celestial cycles. His principles of explanation were in terms of perceptible qualities (e.g., hot or cold, wet or dry) and a series of causes (e.g., matter, agent, plan and purpose). His writings were the basis of natural philosophy (as science was called in earlier days) up to the 17th century. Aristotle was centuries ahead of his times, and has remained a source of insight and instruction till today.

Greek culture flourished in the empire created by Alexander the Great. He patronised famous scholars. Some of them established centres of learning like a famous Museum at the planned city of Alexandria. These centres had large libraries and lent continuity to research work by preserving records and instruments. Although this period, roughly 323 B.C. to 30 B.C., did not reach the heights of the earlier age, it produced some great mathematicians like Euclid, Archimedes and Appoloniu and astronomers like Hipprchus. Studies in medicine and physiology also advanced during this period.

5.1.4 Science in Other Civilisations

(A) Islamic Culture: To European science, Islamic culture is the most relevant one. In the first place, Islamic religion itself is very closely related to Judaism and Christianity, the two main religions of Europe. Secondly, there was an active cultural contact between the Arabic-speaking lands and Latin Europe. Ironically, the great age of Islamic culture coincided with the declining period of culture in western Europe. Drawing on the tradition of Greek science, the early Arabic rulers of Baghdad in the ninth century had the bulk of treaties on Greek science translated into Arabic. Soon after, their own scholars advanced further, especially in mathematics, astronomy, optics, chemistry and medicine. In the Islamic culture, however, the social base of science was rather thin. In a theocratic society, medicine alone of all the sciences was acceptable by the society at large. As a result, no single centre of science flourished for very long. This loss of continuity prevented a sustained development of science in the Islamic culture. The great men of science in the Islamic culture could make creative contribution to science individually, but cooperative scholarship and collaborative efforts were rare. Contact between Islam and Latin Europe came mainly through Spain, where Christians and Jews acted as intermediaries and translators. The twelfth century saw a heroic programme of translation of the work from Arabic to Latin. Despite its leadership, the Islamic civilization was under pressure from barbarians all along its periphery; so it soon went into decline. However, the Islamic culture rendered enormous service to the Western civilization by preserving and then transmitting to it the Greek heritage. In addition, Arab also contributed to modern science a number of words mainly related to plants and foods and also terms like "alcohol" and "algebra".

(B) India: The Indian civilization is an old civilization still alive. It achieved a relatively high level of science and technology at an early stage of its development. Europeans came in contact with the scientific development in ancient India mainly through the Arabic sources. Indian mathematics, with a highly developed system of numeration and reckoning, had a profound influence on Arabic algebra. It also provided the principal Arabic

numerals, 0 to 9. The characteristic science of the ancient Indian civilisation was that of the higher human consciousness. The scientific achievement of Europe and India in ancient times cannot be compared strictly, but should be regarded as being complementary.

(C) China and Japan: China offers more challenges to the historian of European science than India. The reason is that, in spite of long distances between them and their totally different languages, there was continuous contact between Europe and China since the time of Greek civilisation. Usually, this contact was indirect and restricted to the trade in luxury goods; but there was considerable personal contact in the thirteenth century. Marco Polo is the most famous example of this personal contact. Chinese technology until the Renaissance was consistently more advanced than European technology. The British historian of science, Joseph Needham, has shown the patterns of transmission of a series of important inventions from China to Europe. Indeed, the three great inventions (magnetic compass, printing press and gun powder) that were crucial for the transformation of European society came from China. However, Europe never acknowledged its debt to China in the field of science and technology. The Chinese, on the other hand, never achieved the breakthrough to modern science of the European type.

Japan was a cultural colony of China for many centuries. Japan had a brief exposure to Western science and religion before its rulers decided in the seventeenth century to close the doors on such alien and dangerous influence. In the later part of the nineteenth century, the Japanese decided to assimilate with the rest of the world and they did so with vengeance. Their own native religion was vague enough to easily accommodate any assertions of Western science. The result was that Japan rapidly mastered Western science and technology and became one of the leading nations of the world. At present, the Japanese scientists, technologists and common people manage to live partly in a super modern world of high technology and partly still in a world of ancient and rigid social tradition.

5.1.5 Creation of European Science

The "science", on which histories are written, is essentially European. Natural science is a distinctive creation of Europe and its cultural colonies. The creation of European science occurred in two phases; the technical developments in the sixteenth century and the philosophical revolution in the seventeenth century. Out of these came the idea of "science" that is current to this day. The printed books of the sixteenth century in their modern reproduction provide a convenient source of evidence for the state of science during that period. At the beginning of the sixteenth century, scientific knowledge was still rudimentary and largely dependent on ancient Greek and Arabic sources. By the mid-sixteenth century, there appeared scientific works that surpassed the best of their predecessors. In astronomy, *De Revolution bus* was published in 1543 by the Polish astronomer Nicolaus Copercus, which was a technical masterpiece and a revolutionary treatise on cosmology. Similarly, in mathematics, the Italian mathematician Gerolamo Cardano advanced algebra by providing general solution of the cubic equations in his treatise *Arsmagna* published in 1545. During the sixteenth century, the Protestant Reformation caused a series of wars for which military officers needed certain new mathematical skills related to fortification and gunnery. The wars also called for new classes of professionals such as military surgeons and engineers. As a result, there was an enormous improvement in all these arts. By the end of the sixteenth century, applied mathematical arts had become a standard part of the education of a gentleman on the Continent. This was crucial for the formation of new philosophy and its acceptance by the liberally educated audience to which it was directed. In the seventeenth century, there was a radical recasting of the objects, methods and functions of natural knowledge. The new objects of investigation were the regular natural phenomena in a world devoid of human and spiritual properties. The new methods were of disciplined cooperative research, while the new functions of study were the fruitful combination of scholarly knowledge and industrial power. Although this is frequently called the "scientific revolution", it was in fact a revolution about science, rather than one in science itself. The chief architects of the seventeenth century scientific revolution were Francis Bacon in England, Rene Descartes in France and Galileo in Italy. All of them were committed to a great mission, over and above particular facts and theories. Bacon's contribution to actual science was nil, but he provided inspiring ideals and shrewd judgments on the social activity of science. Descartes created a new metaphysics, a radically improved algebra and geometry and some results in physics (the explanation of the rainbow). Galileo, on the other hand, brought clarity to the science of motion, by his studies in mechanics, and laid firm foundation for future work. In spite of their differences in style and contributions, these three prophets of science shared a common commitment about the natural world and its study. Nature itself was seen by them as devoid of spiritual and human properties and had to be studied, therefore, impersonally by using sense experience and reason. The Royal Society in London was chartered in 1662 by the newly restored king Charles II. The French followed the suit a few years later, and other continental monarchs did the same over the next few years. Sir Isaac Newton, one of the greatest scientists and mathematicians of all time, brought the heaven and the earth together in his law of universal gravitation formulated towards the end of the seventeenth century. His masterpiece *Principia Mathematica,* published in 1687, was in many ways the culmination of the scientific revolution in the seventeenth century. The image of Newton dominated science in the eighteenth century. The early eighteenth century, however, was a complacent age. The science that was practised during this period was mainly consolidation. A handful of great mathematicians like Bernoulli developed further the differential and integral calculus invented earlier by Newton in England and Leibnitz in Germany. A successful theory of static electricity was established by the American scientist and diplomat Benjamin Franklin. Although these and a few other similar achievements did not comprise a great science, they firmly established a certain style of science-the one advocated by the prophets of the new philosophy (Bacon, Descartes and Galileo).

Towards the end of the eighteenth century, there began an Industrial Revolution that transformed Europe from an agrarian to an urban society. The activity of science experienced analogous changes. It was during this period that social and institutional foundations were laid for the maturing of science in the nineteenth century. At the beginning of this period, science was a very small-scale activity, pursued as a hobby mainly by interested gentlemen of means and by trained professionals like physicians and engineers. Only a very few universities like Edinburgh in England and Leiden in Holland provided effective instruction in science. The mathematical sciences were well developed, but physics was still a scattering of experiments with qualitative and speculative theories. Chemistry was entirely empirical, while the life sciences were concerned mainly with collectors' activities. The contribution of the Industrial Revolution to science was at first indirect. In the industrialised regions of Britain, there developed an audience for scientific results. By the end of the eighteenth century, free-lance lecturing on science subjects and publication of scientific journals for specialists became economically feasible. On the continent, the more progressive monarchs established specialized colleges of engineering (industrial, civil or military), which provided training for potential recruits.

Starting somewhat earlier than the Industrial Revolution was a movement in France that brought science into the realm of politics for the first time. Called the *Enlightenment,* it had as its programme the struggle against church dogmas and popular superstitions, and it used the facts of science and its rational methods as the main weapon. This movement, started in the 1730s by Voltaire, led ultimately to the French Revolution. National science was deeply involved in the French Revolution. From the Enlightenment, the revolutionaries inherited a faith in science and its methods that permitted the scientists to devote themselves to the organization of war industry for the defence of the Republic in its hour of need. During and after the French Revolution, there appeared a state-supported system of education, with scholarships for talented studen⁺

subsidies for research and rewards for invention. One lasting product of this Revolutionary Science was the *Metric System,* which was intended to produce a coherent system of measures based on the natural units and the decimal system of numbers.

5.1.6 Science in the Nineteenth Century

During the nineteenth century, the industrially advanced nations of Europe assimilated the consequences of the Industrial and French Revolutions. Modern urban society developed in one nation after another with a constantly improving industrial base. In one scientific discipline after another, there was a similar steady progress in the creation of coherent system of scientific institutions. Science as a whole shared the general optimism of the age, receiving credit for its contribution to industrial progress. In retrospect, the nineteenth century appeared as a golden age of matured science. During this century, science expanded successfully into new fields of inquiry, including a combination of mathematics and experimental physics, the application of theory to experiments in chemistry, and controlled experimentation in biology. This was greatly facilitated by the establishment of new and reformed universities in which research was fostered. National and international meetings, for both general science and its specialised areas, became common. The principle of socially organized research, rather than inquiries by isolated talented individuals, became effective during this era. During the nineteenth century, each major branch of experimental science made a great progress. Physics achieved closed union of precise experimentation with abstract mathematical theory that brought unprecedented depth of knowledge and great power in application of that knowledge. Different fields of physics were brought under control and then successively unified by the concept of energy. Thermodynamics united the sciences of heat and work, and then enabled development of a theory of chemical change. Electricity and magnetism were united, first experimentally and then theoretically, by Orsted in Denmark and Faraday and Maxwell in Britain. The general properties of matter were successively mastered and made coherent. Chemistry could be built on the theoretical foundations of Lavoisier's nomenclature and Dalton's atomic theory. Several decades were spent on the heroic task of classifying substances into elements and compounds. The 1860s saw three great victories in chemistry. In Italy, Cannizzaro solved the twin puzzles of atomic weight and chemical composition by the synthesis of neglected earlier ideas (especially Avogadro's hypothesis) and new experimental results. Kekûle in Germany discovered the true structure of organic compounds, with the alternate bonds of the benzene ring. Meyer in Germany and Mendeleyev in Russia mastered the structure of the Periodic Table of chemical elements, and could predict the properties of unknown elements. Thereafter, chemistry could move towards a closer unity with physics, and could gain an increasing power in industrial applications. In biology, the experimental approach was first successfully developed in physiology, mainly by the school of Johannes Mûeller in Germany. The French, who tended to believe in the special character of vital forces, advanced more synthetic aspects of physiology and medicine. Notable among French biologists was Claude Bernard in medicine. Through the field-work in biology and geology came perhaps the most important conceptual achievement of the nineteenth century, viz., the time dimension in nature. The geological record of fossils showed evidence of such complex series of transformations of structure and so many successions of life-forms that the attempts to reconcile this history with that mentioned in the Bible became increasingly difficult. In England, the naturalists Charles Darwin and Alfred Russell Wallace conceived of natural selection by the principle of survival of the fittest. The doctrine of evolution, formulated by Darwin in 1859, unified all these historical disciplines.

The theme of this age was progress, and science received credit for much of the real progress that occurred. Three factors were present in the general praise of science. First came the ancient tradition of respect for learning as a contribution to civilisation. Secondly, a more popular appeal was based on the application of science, first to industry and later to medicine. A third factor was inherited from the Enlightenment, viz., natural science as a weapon against religious dogma and popular superstition. Taken together, these three beliefs remained a strong inspiration for science until very recently.

5.1.7 Science in the Early Twentieth Century

Certain tendencies of science during the nineteenth century became heightened during the early twentieth century. At this time, science was professional in its social organisation and positive in spirit. It was conceived as essentially the work of pure research, with teaching as an ancillary activity and industrial application as a task for other workers in other institutions. Almost all researches were done by highly trained experts, employed mainly in special institutions. Communities of scientists, organized by disciplines and by nationality, enjoyed a high degree of autonomy in setting of goals and standards of research, and in employment and rewarding their members. Individual scientists tended to become very specialised research workers as a result of keen competition. As the cost of research increased beyond the means of individual scientists, science depended on grants from large agencies. The dominant style of work in the first-half of the twentieth century was *reductionist*. Scientific investigations were concentrated on pure, stable and controllable processes achieved in the laboratory. The favourite theories during this era were those involving the simplest physical causes and using highly mathematical arguments. The positive spirit of science was shown by its increasing separation from philosophical reflections. Although Einstein's theory of relativity in 1905 and Heisenberg's "uncertainty principle" in 1927 did raise vigorous philosophical discussions among scientists and laymen alike, these were exceptions.

The scientific achievements during the early twentieth century are too numerous to list. In each major field, progress was based on the successful descriptive work done in the nineteenth century. The early part of the twentieth century saw work in new areas like X-rays, radioactivity and atomic theory. These required a recasting of the fundamental laws of physics in the form of relativity and quantum theory. Chemical methods of analysis were necessary for much of the work in physics. Conversely, the new physical theories (the quantum theory, in particular) were sufficiently powerful to provide effective explanation for a wide variety of chemical phenomena. On this basis, the chemical industry could produce an enormous range of synthetic substances like fibres and plastics. In life sciences, chemical and physical methods brought discovery and explanation of many subtle agencies like vitamins and hormones and the reconstruction of complex cycles of chemical transformations in the living matter. Medical science could build on bacteriology and, through the discovery of specific and general drugs (like Salvarsan, Sulfonamides and Penicillin) which could eliminate both the classic epidemic diseases and also the cruel diseases of childhood. It is no wonder that the triumph of science seemed to promise knowledge and power in superabundance. At present the intimate connection of science with industry, defence and politics has rendered the ideal of pure science obsolete and has confronted scientists with the need for a working conception of the natural world different from the reductionist model of the theoretical physicist. Modern science is an integral part of civilization and, therefore, the survival of science and the survival of civilization go together.

5.2 HISTORICAL DEVELOPMENT OF TECHNOLOGY

5.2.1 Origin and Nature of Technology

The systematic study of technology as a special branch of human activity is a fairly recent phenomenon, in spite of its roots extending back to classical Greece. The word "technology" itself is a combination of two Greek words, viz., *techne* (meaning art or craft) and *logos* (meaning word, speech or knowledge). Thus, the term "technology" in Greek means a discourse on the arts, both fine and applied. When the word "technology" first appeared in English in the seventeenth century, it meant discussion of the applied arts only. By the early twentieth century, the term was coming into general usage and embraced a growing range of means, processes and ideas in addition to tools and machines. By the second half of the twentieth century, technology came to be defined as "the means or activity by which man seeks to change or manipulate his environment".

5.2.2 Science and Technology

Until quite recently, science remained largely divorced from technology, with each pursuing its separate path and maintaining its separate identity. Science in the classical world belonged to the aristocratic philosophers and embodied knowledge, while technology was the possession of the working craftsmen. This separation was due to the fact that the speculative science of Aristotle and other philosophers had little relevance for the practical problems of tanners, millers, silversmiths and other craftsmen. It was not until the lively social-economic interchange, provoked by the commercial revolution of the Middle Age, that science and technology began to draw closer together. The robust growth of technology during this period, involving improvement in sailing ships, waterwheels, windmills and firearms, could not fail to attract the interest and attention of educated men. By the sixteenth century, Francis Bacon was advocating experimental science. The wedding of science and technology proposed by Bacon was slowly consummated. Over the next two hundred years, carpenters and mechanics built iron bridges, steam engines and textile machineries, but it was only during the nineteenth century that technology gradually became science-based. The first scientist who became a major figure in technology was Justus von Liebig, the German father of organic chemistry and the inventor of synthetic fertilizer. The great nineteenth century inventors built on the work of scientists. Edison (the inventor of the first practical system of electrical lighting) built on the scientific work of Faraday and Henry; Bell (the inventor of telephone) built on the work of Helmhõltz, while Marconi (the inventor of radiotelegraphy) built on the earlier work of Hertz and Maxwell. Massive trial and error process by which Edison discovered carbon filament for his electric light bulb resulted in creation of the world's first research laboratory (outside the university system) at Menlo Park, New Jersey (U.S.A). Thus, the date of Edison's demonstration of electric light (October 21, 1879) may also be considered as the birthday of modern technological research. From this point onwards, application of scientific knowledge and scientific principles grew at an accelerated rate. The engineering rationalism that Frederick W. Taylor applied to the organisation of workers in mass production, and the time and motion studies of Frank and Lillian Gilberth, were logically followed later in the twentieth century by the invention of systems engineering, operations research, simulation studies and mathematical modeling. These developments, added to the increased specialization and professionalisation of technological work, brought technology to its highly efficient present state.

5.2.3   Social Involvement in Technology

Technology may be defined as "the systematic study of techniques for making and doing things". Thus the history of technology is, in a sense, the history of man. Man may not possess highly developed instinctive reactions, but does have the capacity to think creatively and systematically about techniques. By virtue of his nature as a tool maker, man may be regarded as a technologist from the very beginning, and the history of technology encompasses the whole evolution of man. The history of technology reveals a profound interaction between the incentives and opportunities of technological innovation, on the one hand and the sociocultural conditions of the human society within which they occur, on the other hand. There are three points at which there must be some social involvements in technological innovation viz., social needs, social resources, and a sympathetic social environment. In the absence of any one of these three factors, it is unlikely that a technological innovation will be widely adopted. The sense of social needs must be strong, or people will not be prepared to devote financial and other resources to a technological innovation. The needs may be for a more efficient cutting tool, a more powerful lifting device, a labour-saving machine, a means of utilizing new fuels, or a new source of energy. Military needs have always provided a stimulus to technological innovation. In modern society, needs have been generated by advertising.

Social resources are similarly indispensable prerequisites to a successful technological innovation. Many important inventions were unsuccessful because the social resources vital for their realization (capital, materials, skilled personnel, etc) were not available at the time of the invention. The note books of Leonardo da Vinci are full of ideas for helicopters, submarines and aeroplanes, but none of these reached even the stage of working models because of the deficient resources of one sort or the other in that era. The resource of capital involves the existence of surplus productivity and an organization capable of directing the available wealth into channels in which the inventor can use it. Material resources involve the availability of appropriate metallurgical, ceramic, plastic or textile substances that can perform whatever functions a new invention requires of them. The resources of skilled personnel implies the presence of technical skill capable of constructing new artifacts and devising novel processes. A society, in short, has to be well primed with suitable resources in order to sustain technological innovation. Whatever the psychological basis of inventive genius, there is no doubt that the existence of socially important groups willing to encourage inventors and to use their ideas has been a crucial factor in the history of technology.

5.2.4 Chronological Developments in Technology

The material for the study of history of technology comes from a variety of sources. For many millennia of unrecorded history, in which slow but steady and substantial technological advances were made, it is necessary to rely heavily upon the archaeological evidences. Various chronological phases of the historical development of technology are presented in Table 5.2.4.1.

Table 5.2.4.1 Chronological Development of Technology

Chronological Phase	Major Technological Developments
The Stone Age (Up to 3000 B.C.)	Earliest communities, The Neolithic Revolution.
The Urban Revolution (3000-500 B.C.)	Crafts, use of copper and bronze, irrigation, urban manufacturing, building and transmitting knowledge.
Greek Civilisation (500 B.C.- 500 A.D.)	Mastery of iron, mechanical contrivances, agriculture and building.
Middle Age (500- 1500 A.D.)	Power sources, agriculture and crafts, architecture, military technology, transport and communications.
Emergence of European Technology (1500 – 1750 A.D.)	The Renaissance, steam engine, metallurgy and mining, new commodities, agriculture, construction, transport, communications and chemistry.
Industrial Revolution (1750 – 1900 A.D.)	Windmills, steam engines, electricity, internal combustion engines, petroleum, metallurgy, mechanical engineering, textiles, chemicals, agriculture, civil engineering, transport, communications and military technology.
The Twentieth Century (1900 – 1945)	Fuel and power, industry, food and agriculture, civil engineering, transportation, communications, military technology, atomic energy, aeroplanes, helicopters and electronics.
The Twentieth Century (1945 – 2000)	Power, new materials, automation, computers, information technology, space technology, food production, biotechnology, transport, communications, military technology, nuclear technology, population control and environmental technology.

5.2.5 Technology in the Twentieth Century

It is difficult to write the history of technology in the twentieth century because of the large mass of the materials as well as the problem of distinguishing the significant from the insignificant events. Inspite of immense achievements in technology till the end of the nineteenth century, the twentieth century witnessed more advances in technology over a wide range of activities. The aeroplane, rockets, interplanetary probes, electronics, atomic energy, antibiotics, insecticides, computers and a host of new materials have all been invented and developed to create an unparallel social situation, full of great possibilities and even greater dangers, which were virtually unimaginable in the nineteenth century. From the point of view of the history of technology, it is convenient to divide the twentieth century into two periods, viz., 1900–1945 and 1945–2000. The years 1900 to 1945 were dominated by the two world wars, while those since 1945 were preoccupied by the need to avoid major wars involving the possible use of nuclear or biological weapons. The year 1945, dividing these two periods, is one of outstanding social and technological significance. It was in 1945 that the first atomic bomb was detonated in New Mexico (U.S.A) in July and then two such bombs were dropped on the towns of Hiroshima and Nagasaki in Japan in August.

There have been profound political changes in the twentieth century related to technological capacity and leadership. It might be an exaggeration to regard the twentieth century as the "American century". The rise of the US has been based upon tremendous natural resources exploited to secure an increased productivity through widespread industrialisation. This objective was tested and demonstrated in the two World Wars. In course of these wars, the technological leadership passed from Britain and other European countries to the US. The two wars were themselves the most important instruments of technological as well as political changes in the twentieth century. The rapid evolution of the aeroplane is a striking evolution of this process. The dramatic appearance of the tank in the First World War and that of the atom bomb in the Second World War show the same sign of technological response to an urgent military stimulus. The same can be said for the radar, the computer, operations research, games theory and systems analysis which were all developed during the Second World War. The Wars were responsible for accelerating the transformation from "little science" with research still largely restricted to small-scale efforts by a few devoted but isolated scientists, to "big science" with emphasis on large multi-disciplinary research teams sponsored by governments and large corporations, working collectively in the development and application of new techniques. There is little doubt that the change in the scale of technological enterprises has far-reaching consequences. This has been one of the most momentus transformations of the twentieth century since it has altered the quality of industrial and social organisation. In the process, this transformation has assured technology a position of importance and honour in the society, and this has happened for the first time in the long history of technology.

5.3 APPROPRIATE TECHNOLOGY AND CRITERIA FOR ITS ASSESSMENT

5.3.1 Appropriate Technology

The research and development resources of the world have increased enormously after the Second World War. In most countries the main objectives, which have justified the large research and development expenditure, have been defence, issues of national prestige (e.g., space research) and economic growth with relatively less efforts in other directions (like improvements in social and service sectors including health and education). In industrialised countries, there is concern with respect to the negative side-effects of technology. Foremost amongst this concern is the environmental pollution. There are many other phenomena like urban deterioration, loss of satisfaction in work and increased alienation of people from society which are mostly the consequences of a technology-dominated culture. All these negative consequences of technology imply a deterioration in the quality of life.

The pursuit of economic growth by the industrialised countries requires still further development of technology. If new technology continues to mushroom without control, environmental pollution and other negative side-effects of technology are likely to go on increasing and as a result, producing great or social and ecological disturbances. Therefore, technology should be evolved to be socially acceptable. This would necessitate a much more careful selection and management of technology. One of the requirements in this regard is to develop the means of **technology assessment** so that one can assess the social, cultural and economic consequences of a particular process. This would avoid unwanted and undesirable surprises at a later stage. This brings in the concept of **appropriate technology**. Appropriateness of technology can only be achieved on the basis of an accepted value system for global human development and interpreted in each country in terms of local economic, social and cultural needs. Appropriate technology is, therefore, a necessary concept for both industrialised and developing countries and not merely an euphemism for persuading the Third World countries to accept and adopt inferior technology. The term "appropriate technology" means the development and utilisation of processes and work organisation most suited to particular circumstances, both economic and social, of a particular country or sector.

For the industrialised countries, present technologies are far from fully appropriate to the present and future needs. In future, these technologies will need to be made more appropriate by taking into account the social concerns, pollution control and conservation of raw materials and energy. So far, Research and Development have been aimed at the creation of technological processes to achieve high level of labour productivity and large-scale production to provide for large markets and thus to take advantage of the economy of scale. Towards this end, the technologies developed in industrialised countries have been quite appropriate. They have, however, failed to take into account the social and environmental factors.

For the Third World countries, "free" technology is available to achieve import substitution in regard to many traditional goods now obtained from abroad. In case of other goods for which no such free technology is available, social and economic benefits would arise by devising small-scale and labour-intensive methods of production, which are capable of being decentralized, thus relieving pressure on urban areas and bringing prosperity to rural areas. In most Third World countries, there is a paramount need to devote more efforts to improving the existing empirical technology of the traditional sector, which is generally neglected. So far, many of the developing countries have tried to replace the age-old traditional methods and tools of production by imported technology, often accompanied with social resistance, inefficient operation and uncertain success. Imported technologies are not always seen as conforming to local needs and cultural habits. Much can be done, therefore, by the application of simple and well-understood scientific principles to the improvement of the traditional tools and methods, with the possibility of greater improvement in efficiency and minimisation of cultural disturbances.

The technological needs of most Third World countries, although exceedingly diverse, fall into five major categories:

(1) Introduction of science-based technologies, developed elsewhere but modified suitably to take into account the locally available materials, work methods and human skills, within the framework of national economic and social objectives.

(2) Creation of a nucleus of modern industrial development for the overall economic improvement to facilitate future developments.

(3) Introduction of a wide range of established technologies, usually freely available, for the manufacture of basic products presently being imported.

(4) Basic research and development effort, a part of which might be undertaken by co-operation between developing countries with similar needs, to evolve highly efficient but labour-intensive processes.

(5) Construction of a technological service at both national and local levels to improve the existing empirical methods and tools of the traditional sector by the systematic application of scientific principles.

5.3.2 Reduction of Technological Disparities

The prerequisite to reduce economic disparities, both within and between the countries, is the existence of a political will to do so. Economic disparities are intimately linked with technological disparities. It is improbable, therefore, that the economic gap can be bridged without a new approach to the diffusion and assimilation of technology. It is true that the building up of the scientific and technological capacities of the developing countries is an essential element for their accelerated development. However, just as general development requires an influx of capital from rich countries, so it requires the influx of technological capital from industrialised countries for accelerated technological development. This technological capital may be in the form of patents, know-how, marketing expertise, managerial experience, etc. Developing countries should have access to the achievements of modern science and technology. Rich and developed countries should also take steps for promoting the transfer of technology and creation of indigenous technology for the benefit of developing countries. In the first place, the system of science is very different from that of technology. Technological knowledge is not always a free commodity and rather a form of industrial property owned by individuals, corporations or states. Technological know-how is a valuable commodity, which is sold and bought in the national and international markets and, therefore, carefully guarded. Thus, most of the advanced technologies are not automatically available to developing countries. Transfer of technology can be a costly business which, like any other commercial transaction, requires great care in the choice of expensive technologies before purchase. The problem of securing an adequate flow of advanced technology from the industrialised to the developing countries is, therefore, an economic matter.

There are three main requirements for the effective transfer of technology from the industrialised and developed countries to the developing countries:

(1) Skills, on the part of the planning and other agencies of the developing countries, in the selection of technologies which are most appropriate for their economic and social development.

(2) Easy access to and equitable conditions for such transfer of technology.

(3) The building up of an appropriate infrastructure for technology to permit not only wise selection from the available technologies, but also the capacity to modify the processes and products to render them suitable for use with locally available materials and to respond to local cultural conditions.

As it happens, many problems exist with regard to each of the above three needs. For the effective transfer of both processes and products, new attitudes as well as new approaches are required on the part of both donor and recipient countries. Another problem arising with regard to the transfer of technology is that the governments of the market-economy countries are generally not in a position to direct their industries to part with their technologies, which in many cases have been generated at a very high research and development cost and possession of which determines their competitive status in both domestic and foreign markets. Nevertheless, there is possibility for greater improvements in the terms of technology transfer by new means. For example, the donor countries could establish policies which would enable them, as part of their foreign aid policy, to subsidise the sale of technological property to the developing countries. Special arrangements have to be made between the industrialised countries with regard to such transactions where multinational corporations are involved. Within the receiving countries, on the other hand, the prime necessity is to establish deliberate polices for technological development. These policies should be related organically to their long-term economic and social objectives. It is extremely important that there should be a capacity in the developing country concerned to select the optimum process. It is all too easy for a developing country, lacking the depth of knowledge of the range of current international scientific and technological trends, to select inappropriate processes and products.

5.3.3 Building National Capacity for Science and Technology

The need for a careful building of an indigenous capacity for research and development within each of the developing countries is the kernel of the development problem. The basic need is to realize that science and technology can only contribute to national development in the long run if they are regarded as inherent and essential elements in a socio-cultural process. A few developing countries like India and Mexico are fully aware of the need for relating industrialisation to the aims of society and of using science and technology intimately coupled with the various sectoral developments; but this is not generally understood by a vast majority of developing countries. There is, however, a growing awareness that each country must create its own scientific and technological capacity to ensure that imported technology takes roots and then spreads. The usual approach is to propose the creation of new universities, to provide more funds for research or to create industrial research laboratories, on the assumption that if there is more research, there will be more technologies and more developments. These are, of course, excellent measures if the new institutions are well conceived and integrated in the socio-economic system of the country concerned. This, however, is not always the case and ill-conceived research institutions breed frustrated scientists, ripe for the brain drain. Too often, universities in developing countries are isolated from the community they are supposed to serve. They become outposts of learning remote from the local problems and prepare individuals of high quality for whom there is no suitable employment. Thus, there is no easy path to the creation of a vigorous scientific and technological capacity. It must be approached simultaneously from many angles; in the universities and technical institutions, in agricultural institutions and extension services and in industry and public services. It must be supported by information services which can scan world developments in science and technology and bring to the attention of the authorities those elements of new knowledge which are significant to the society concerned.

The successful use of foreign technology is, therefore, much more a merely matter of access to foreign patents and know-how, or even the availability of capital to exploit them. It requires the adoption of deliberate policies of technological innovation as part of a long-term economic plan. Some of the factors such a policy should be concerned with are :

(1) Co-operation between the government and industry in acquiring a detailed understanding of the nature of technological innovation.

(2) An explicit statement concerning the long-term national objectives so as to make it possible to analyse the specific needs for new knowledge.

(3) Acceptance of the concept that technological innovation is part of a socio-economic-cultural process and not an autonomous process.

(4) Coupling science and technology with the productive and educational systems of the country to make R & D more effective and self-sustaining.

(5) Developing clear criteria of a social, cultural and economic nature for the selection of imported technology and know-how.

(6) Achieving a balance between the conflicting needs for building highly capital-intensive industries with high economic yields and that for the creation of employment.

(7) Adoption of proper fiscal and tariff policies to facilitate the introduction of new technologies and their protection during the early critical years.

(8) Encouragement of innovation and maintenance of quality standards through government purchases based on specifications of quality.

(9) Forward manpower planning especially for scientific, technological and managerial skills and phasing of educational and training facilities to ensure that these skills are available where and when they are required.

(10) Encouragement of the creation of joint industrial projects with neighbouring countries having similar needs for the purposes of cost sharing, attainment of economies of scale and prevention of waste.

(11) Stimulation of industrial research by subsidy, research contract and tax concessions.

(12) Formulating policies for resource development and management including research on new or improved uses of indigenous materials, both minerals and vegetables, with a view to add value and increase employment.

(13) Provision of information networks for the use of government planners, industrial enterprises and research workers.

(14) Creation of industrial advisory services to assist industrial firms, especially the small-scale industries to achieve the best general practice through efficiency survey, plant layout, work study and better utilisation of fuels and materials.

(15) Institution of public information campaign to inform the citizens, through the educational system and the media, on the nature and objectives of the proposed developments and their full social and economic significance.

The national policy which should follow by taking into account all the above issues can be evolved gradually. Its implementation may be difficult for the smaller developing countries; but such a national policy can serve as a guideline for development. Much international help would be necessary from the United Nations and other agencies, both in the elaboration of general principles and in helping the individual countries.

5.3.4 Assessment Criteria for Appropriate Technology

Compared to earlier decades, we have a higher rate of growth of Indian economy today. In spite of this, we are still facing the chronic problems of poverty and unemployment. The main reason is that we have not used appropriate technology for solving our economic problems. Modern technology and appropriate technology are complementary to each other and not antagonistic. Appropriate technology is not necessarily an inferior or out-dated technology. In fact, appropriate technology is a boon to developing countries since it helps to conserve energy, protects the environment and makes optimum utilisation of the available natural and human resources.

Some of the major criteria for the assessment of appropriate technology are:

(1) The technology concerned should be developed to fulfill specific social needs.

(2) It should use the raw materials, available locally.

(3) It should require the skills, available locally.

(4) The machinery used in the technology should be such that it can be operated and maintained locally or with easily available skills.

(5) The technology should be such that it can contribute to social development in a positive manner.

(6) The products of the technology should be cheap, should have high quality and ready market.

(7) The technology should be less expensive so that it can be utilised by developing countries facing the scarcity of capital.

(8) It should be labour intensive in order to generate employment in developing countries.

(9) It should enable the workers especially in rural areas to utilise their full human potential.

(10) As far as possible, the technology should utilise the traditional skills and capabilities.

(11) Instead of mass production, the technology should encourage production by the mass.

(12) The technology should ensure maximum development with minimum capital outlay.

(13) It should have export potential.

(14) It should make less demand on energy.

(15) The technology should not affect the environment and should have the potential for waste recycling.

There is an urgent need for a concerted global effort to carefully select appropriate technologies of simplest nature and to apply them at the grass roots level with the aim of alleviating poverty and fulfilling basic needs in developing countries. Such a move for the application of appropriate technology to alleviate poverty may be launched parallel to the current efforts for infrastructure building, a development of science and technology policies at the national level.

5.4 TECHNOLOGY ADAPTATION

5.4.1 Some Aspects of Technology Transfer

International transfer of technology is a major aspect of the new international division of labour. In fact, there are four main motivations behind the enormous geographic expansion of the commercialization of technologies. These four motivations are:

(1) The export of technology can be viewed as an attempt to extend the product life cycle of technologies, which is either in a high stage of maturity or is going to become obsolete in the very near future. In this sense, the international transfer of technology fulfils the function of a substitute for innovation. As a result, a major part of the technologies transferred to developing countries consists of mature or obsolete, consumer-goods technologies, which are sold at excessive prices.

(2) The world-wide commercialization has turned out to be a very efficient instrument for the penetration of closed markets. There are three types of growth markets available in developing countries; private luxury goods consumption, government procurement markets and world factory markets. Given the high level of protection surrounding these markets, the export of technology might be the only way to penetrate them. As pointed out by the American industrialist and technology export policy consultant Thomas Callagham, Jr., "Markets closed to products are invariably open to technology".

(3) The international transfer of technology has been perceived by corporate headquarters as a necessary and efficient instrument for shifting the enormous cost burden of R & D activities on the shoulders of weaker bargaining partners. This corporate policy of sharing the burden of R & D among unequals is an important precondition for the present global patterns of technological dominance.

(4) The international transfer of technology has been an essential precondition for the internationalization of production. The massive proliferation of production technologies and strategies of world-wide outsourcing could have been inconceivable without transfer of technology. For the multi-national companies(MNCs), the main problem is to co-ordinate on a global scale the highly complex flow of resources; capital, manpower and technology to secure long-term profitability.

Considering these four aspects of technology transfer, it is obvious that transfer of technology to developing countries is not the result of a kind of international welfare policy. Technology transfer, in fact, is an important element of the strategies of private firms (essentially MNCs), which are forced to internationalise their overall cycle of capital reproduction on a growing scale. Thus, the transfer of technology into developing countries is pursued in such a way that the benefits to be derived from control over the R & D capacities and new technologies can be optimized with regard to markets as well as production sites.

5.4.2 Reform Measures for Technology Transfer

Discussion on a reform of the present system of international transfer of technology has gathered momentum since two decades. Attempt at reform has centred on the development of a technology policy, which could enable governments of developing countries to neutralize the negative aspects of technology imports and, at the same time, increase local innovative capacities. Some important aspects of the reform of the present system of international transfer of technology are:

(1) The international patent system should be modernized in order to remove rigidities of the present system which have become rather

dysfunctional. Modernization can help to avoid unnecessary conflicts and, at the same time, pave the way for codification of new protective mechanisms.

(2) There should be a code of conduct for the transfer of technology, which should make it possible to devise some basic rules of the game for international technology markets. In case of MNCs, who are the main actors in international transfer of technology, such a code of conduct would facilitate long-term planning and help to reduce the high risk involved at present.

(3) Cost disparities, which are hindrance for the transfer of appropriate technology, should be tackled through manipulations of monetary and fiscal policies, and through improved engineering of business cycles at both national and international levels. Such corrective measures should be put into practice in the technology receiver country and also in the country from which the technology originates. Moreover, priority should be given to raw material-intensive technologies, especially to the technologies with a high need for low cost, low skilled, easily trained, and easily displaceable labour.

These reforms should be applied in highly industrialised as well as developing countries. In the industrialised countries, structural readjustment means the sub-ordination of regional structures to technical change in such a way that it will be possible to identify those products and production processes which are suitable for offshore production in developing countries. On the other hand, it is exactly on these products and processes that the policies to strengthen local technological capacities in developing countries should focus.

The main aim of the prevailing reform concepts of the international transfer of technology is limited changes of transfer conditions. What is really required is to attack the technological dominance, i.e., to break open the extremely asymmetric distribution of control over processes of production and innovative capabilities. This necessity has not yet been transformed into politically operational concepts. That is why, the import of technology is being given the key position with regard to policies to overcome technological dependence.

5.4.3 Need for Technology Adaptation

There is now a proliferation of technologies into developing countries. This new proliferation might become instrumental in producing new and qualitatively intensified forms of technological dependence, thus further increasing the economic and political hierarchical structure of the North – South relation. To remedy this situation, it has been suggested that the developed countries should renounce their monopolistic attitudes towards high technology and developing countries should ensure the supply of raw materials to developed countries. In order to reduce the technological dependence, the industrialization of developing countries should proceed along the following lines:

(1) Expansion of certain basic industries like iron and steel, cement, aluminium and petrochemicals.

(2) New pattern of world market subcontracting with regard to certain areas of capital goods production (e.g., production of low-cost machine tools) and engineering consultancy (e.g., world-wide pooling of low-cost, locally subsidized R&D teams).

(3) Attempt to induce the growth of local capital goods and basic industries via the establishment of local small arms industries.

The Department of Industrial Development under the Ministry of Industry, Government of India, maintains a Technology Data Bank which gives information of a wide choice of technologies and collaborators available throughout the world. When a technology has to be imported, one should choose the most appropriate technology available to suit the project in the Indian environment. Our developing economy has a certain level of technology absorption capacity; so it is ideal if one can find a technology suited to that level, with the potential for improvement, keeping pace with the changes in our socio-economic environment. On the other hand, if such an ideally suited technology is not available, we have to import a technology nearest to the ideal one and adapt it to suit our needs.

Certain technologies imported from highly developed countries are usually capital intensive and require much less labour for production. The negative social side-effects of such technologies can be checked by adapting them to suit the local socio-economic conditions and also by applying corrective welfare policies and social engineering measures. The increasing unemployment and marginalization of a majority of the population of developing countries is the result of job-destructive character of the imported, capital-intensive technologies. This has to be mitigated by planned adaptation of such technologies to suit the local needs.

The mechanism of technology transfer is such that the productive system of a developing country acquires a technology developed in a highly industrialized country. To the technology exporter, the benefit ranges from international market expansion to higher profits from joint ventures and licensing fees. For the recipient country, there are the obvious benefits of developing a stronger technological base and contributing to the process of economic growth. In addition, there is also the initial resource-saving benefit of not having to develop the technology concerned indigenously, but to import it from the country where it has been developed, debugged and tested. When such a technology is imposed on a different socio-economic base of a developing country, it may lead to many undesirable consequences. It may result in an enhancement of social and economic inequalities, greater dislocations and more concentration of economic power. Being capital intensive, the imported technologies may tend to intensify the problem of resource misallocation. Such technologies may also tend to be environmentally destructive to the developing economy. Due to the monopolistic practices of multinational companies, the developing economy incurs a heavy burden of foreign exchange to pay for the direct and indirect cost of technology transfer. Moreover, there is an adverse social impact of reduced pressure to learn and perform in developing countries if they can easily get a pre-packaged technology on a turn-key basis.

The need for technology adaptation becomes obvious when we consider the fact that the technology developed in a highly industrialized country is, in most cases, capital-intensive as well as energy-intensive. These traits are not consistent with many of the recipient countries' natural and other resources. Furthermore, the technology developed in the exporting countries is often designed to meet an artificially stimulated demand for luxury goods in a high-consumption society. Such a technology does not exactly meet the needs of a developing country. In addition, since the imported technology is usually capital and energy-intensive, it often has adverse environmental and ecological impacts. In fact, the developed countries carry out technology transfer to serve their own intrinsic objectives, rather than to serve the needs of developing countries. Due to all these factors, it follows that the developing countries have to make a very careful selection of the technology to be imported, and then adapt it to serve their own unique socio-economic objectives.

5.4.4 Some Strategies for Technology Adaptation

In case of technology transfer, the intrinsic objectives of the technology-exporting industrialized countries include acquisition of a wider mono-polistic production base, expansion of their profit-making operations, exploitation of cheap labour in developing countries, and to escape rigid environmental and occupational-safety legislations in their own countries. It is in view of these objectives of the technology exporters, and the lack thereof in the developing countries, that the issue of technology selection and its adaptation to suit local needs assume critical importance. Developing countries have to be very selective in the type of technology they choose to meet their prescribed objectives and criteria. The technology should be selected with a view to enhancing their resource base, to suit their socio-economic conditions, and to be consistent with their natural, capital and human resources. The technology to be imported should also meet certain environmental constraints, promote self-reliant development and indigenous research capability, and lessen technological dependency in future. It is obvious that very few technologies developed in highly industrialized nations can meet most of these criteria; so it is the responsibility of the developing countries to adapt the imported technologies to satisfy their requirements.

Once a national development plan is formulated in a developing country, it should also carry with it a national technology plan. Decisions have to be made as to whether the technological inputs into national development efforts should be imported (and, if necessary, adapted to meet local requirements), or be met from local sources (e.g., research institutions and universities in the developing country itself). The technological plan may call for establishment of an institutional infrastructure (such as a national technology centre), with the objectives of selecting appropriate technologies for import, adapting them to suit local conditions, and developing appropriate domestic technologies to substitute the imported ones. With a technological plan designed within the broad framework of the national development plan, a set of criteria and guidelines should be developed for selecting the technologies to be imported, adapting them to match the requirements of national development objectives, and deciding the mode of their implementation (such as joint ventures, licensing contracts, turn-key plants, etc.).

For efficient adaptation of imported technology, the local technological infrastructure, especially with regard to information networks, communications systems, standardization, and adaptative activities should be developed sufficiently. These should be achieved by modernization of the educational system on one hand, and establishment of science and technology institutions, on the other hand.

The job-destructive character of imported technologies should be mitigated by proper adaptation to make them more labour-intensive and less capital-intensive. This is especially important in case of imported technologies related to those parts of the agricultural sector, which have not yet become fully integrated into capitalistic production, subsistence activities, and subcontracting activities of small firms. The growing poverty of a majority of the world population, especially the unimaginable increase of rural poverty, should be countered by a basic human needs strategy. Such a strategy should focus on the development, diffusion and adaptation of technologies in the fields like food production, habitat, health-care and elementary education. Such technologies, whether imported, adapted or locally developed, should be cheap, easy to transfer, easy to learn, and easy to maintain. No one can deny the urgency of developing and adapting such technologies on a priority basis.

Another important objective of technology adaptation is to enhance the absorptive capacity of local enterprises of small and medium size. This can be achieved by establishing complementary scientific and technological infrastructures in developing countries for the purpose of worldwide screening and tapping of new scientific and technological developments and adapting them for local needs. The developing countries should also adapt imported technologies to take full advantage of their cheap, disciplined, easily displaceable and highly qualified labour, especially the scientific and technical professionals, laboratory technicians, consultancy firms, and business lawyers.

It is within a society that the technological needs have to be defined with regard to the optimal use of local resources, and the utmost fulfillment of basic human needs. Once a society or a developing country defines its technological needs, the decision as to which technology to import and how to adapt it to satisfy the local socio-economic needs, becomes easier.

MODEL QUESTIONS

(Essay/Long Type)

5.1 Discuss the status of Science during the 20th century.

5.2 Describe the technological developments during the 20th century.

5.3 Define and explain "appropriate technology".

5.4 List the major criteria for assessment of appropriate technology.

5.5 Write a critical note on technology adaptation.

(Objective/Short Type)

5.6 Science was fairly established :
   (a) Before the 10th century
   (b) During the 14th century
   (c) During the 20th century
   (d) With effect from the 17th century.

5.7 Industrial Revolution transformed Europe from :
   (a) Agrarian to urban society
   (b) Capitalistic to socialistic society
   (c) Scientific to religious society
   (d) None of these.

5.8 A major outcome of the French Revolution was :
   (a) Mixed economy   (b) Metric system
   (c) Both   (d) Neither.

5.9 The early part of the 20th century saw work in areas like :
   (a) X-ray   (b) Radioactivity
   (c) Atomic theory   (d) All of these.

5.10 Match the following :

(A)	Plato	(a)	Greece
(B)	Aristotle	(b)	Italy
(C)	Marco Polo	(c)	Republic
(D)	Galileo	(d)	China

5.11 Write true or false:

Justus von Liebig was never a major figure in technology.

5.12 Write true or false:

The 20th century may be regarded as the American century.

5.13 Write true or false:

Technological know-how is rarely a priced commodity.

5.14 Write true or false:

For adaptation of imported technology, the local technological infra-structure should be sufficiently developed.

5.15 Write true or false:

Usually, technologies imported from highly developed countries are not capital intensive.

5.16 Write the odd ones:

(a)	Locally available skills	(b)	Less expensive product
(c)	Export potential	(d)	Capital intensive.

5.17 Fill up the blank:

_____ in Germany discovered the true structure of organic compounds.

5.18 Fill up the blank:

The doctrine of evolution was formulated by _____.

GROUP - II

Environment

Environment

6.1 INTRODUCTION

Human beings live in the realm of nature, they are constantly surround by it and interact with it. The most intimate part of the nature in relation to man is the biosphere. Man is aware of the influence of nature in the form of air he breathes, water he drinks and food he eats. Mankind and civilization cannot exist without these precious gifts of nature. Earth is the only planet where other living beings have evolved because of the availability of all the ingredients such as air, water, food, energy, etc., which are essential for life. These together constitute the environment. The environment can be defined as one's *surroundings*. To the environmental engineer, the word *"environment"* may take on global dimensions, may refer to a very localised area in which a specific problem has to be solved, or may refer to a small volume of solid, liquid or gaseous materials. The importance of environment has been underestimated for too long, resulting in damage to human health, reduction in productivity and undermining the future development prospects. After a century of swallowing noxious fumes and toxic chemicals released by automobiles, industries, power plants, etc., mother nature has begun to show sign of sickness and serious environmental damage.

6.2 THE BIOSPHERE

The global environment consists of atmosphere, hydrosphere and lithosphere in which the life-sustaining resources of the earth are contained. The *atmosphere*, which is a mixture of various gases (mainly nitrogen, oxygen, carbon dioxide and water vapour) extending outward from the surface of the earth, evolved from the elements of the earth that were gasified during its formation. The *hydrosphere* consists of oceans, lakes and streams and shallow ground water bodies. The *lithosphere*, on the other hand, is the soil mantle that wraps the core of the earth. The *biosphere*, a thin layer that encapsulates the earth, is made up of atmosphere and lithosphere adjacent to the surface of the earth, together with hydrosphere. It is within the biosphere that all the life–forms of earth live. Life-sustaining materials in solid, liquid and gaseous forms are cycled through biosphere, giving sustenance to all the living organisms. Life-sustaining resources, air, water and food, are withdrawn from the biosphere. It is also into the biosphere that waste products in gaseous, liquid and solid forms are discharged. From the beginning, the biosphere has received and assimilated the wastes generated by plant and animal life. Natural systems have been always active, dispersing smoke from forest fires, diluting animal wastes washed into streams and rivers and converting debris of past generations of plant and animal life into soil, rich enough to support future populations. Though the natural system has operated for millions of years, it has now begun to show signs of stress, primarily because of the severe impact of man upon the environment.

6.3 MAN'S IMPACT UPON THE ENVIRONMENT

6.3.1 Environmental Equilibrium

Various life-forms of the earth live in natural equilibrium with their environment. Natural and manufactured wastes generated and released into the biosphere by the increased number of human beings have upset the natural equilibrium of the environment. Anthropogenic (human-induced) pollutants have overloaded the ecological system. This overloading has come relatively late in the course of human interaction with the environment since the early societies were primarily concerned with meeting their natural needs, when human shared his needs with most of the higher mammals. The people belonging to early societies were not concerned themselves with meeting the acquired needs associated with more advanced civilizations.

6.3.2 Satisfying Natural Needs

Early human beings used natural resources (air, water, food and shelter) mainly to satisfy their needs. The residues generated by the use of these resources were generally compatible with the environment, or were readily assimilated by it. Primitive humans usually ate raw plant and raw animal without disturbing the atmosphere with the smoke from fire. When the use of fire became common for cooking and warmth, the small amount of smoke generated by these activities were rapidly dispersed and easily assimilated by the atmosphere. Early civilizations drank from the same rivers and streams in which they bathed and deposited their wastes. However, the impact of such use of natural water resources was relatively low since natural cleansing mechanisms (self-purification processes) easily restored the water quality. The early humans used natural shelters like caves, or built their homes from wood, mud or animal skins. They left behind items that were readily decomposed by micro-organisms and absorbed by the atmosphere, hydrosphere, or lithosphere. After early people began to settle down in large groups, their impact upon their local environment began to show significant adverse effects. By the late eighteenth century, the water of the rivers Rhine in Germany and Thames in England became so polluted that they could not support game fish.

6.3.3 Satisfying Acquired Needs

With the dawn of the Industrial Revolution, humans were better able than ever before to satisfy their natural needs of air, water, food and shelter. By the early twentieth century, automobiles, appliances, processed food and beverage became popular as necessities. Meeting these acquired needs became a major thrust of the modern industrial society.

Unlike natural needs the acquired needs are usually met by products that must be processed or manufactured or refined. The production, distribution and use of such products usually result in more complex residues, many of which are not compatible with the environment, or readily assimilated by it. As a general rule, meeting of the acquired needs of modern societies generates more residuals than meeting the natural needs. These residuals are likely to be less compatible with the environment and less likely to be readily assimilated into the biosphere. As modern societies ascend the socio-economic ladder, the list of acquired needs or luxuries increases and so do the complexities of the production chain and the amounts and varieties of the pollutants generated. As a result, the impact of modern human populations upon the environment is of major concern to the environmental engineers and scientists.

6.4 ENVIRONMENTAL POLLUTION

6.4.1 Pollution and Pollutants

The word *"pollution"* is derived from the Latin word *polluere,* which means to contaminate any feature of the environment. Environmental pollution may be broadly defined as "adding to the environment a potentially hazardous substance or source of energy faster than the rate at which the environment can accommodate it". Alternately, environmental pollution may also be defined as :

> "An undesirable change in physical, chemical or biological characteristics of air, water and land that may be or will be harmful to human life and other living organisms, living conditions, industrial progress and cultural assets and will deteriorate raw material resources".

In brief, one may define the environmental pollution as "an unfavourable alteration of environment largely as a result of human activities".

A pollutant is a substance that affects adversely or alters the environment by changing the growth rate of species, interferes with the food chain, health, comfort and amenities of the people and is toxic. In the Environment Protection Act (1986), passed by the Government of India, an environmental pollutant is defined as :

> "Any solid, liquid or gaseous substance present in such con- centration as may be or tends to be injurious to the environment and the environmental pollution means the presence in the environment of any environmental pollutant".

Environmental pollutants may broadly be divided into two categories : (1) biodegradable and (2) non-biodegradable. Natural processes involving micro-organisms easily decompose the biodegradable pollutants. Domestic sewage comes under this category. The biodegradable pollutants cause environmental pollution only when their production exceeds the capacity of the environment to degrade them. Action of microbes like bacteria, fungi and protozoa in a sewage treatment plant enhances nature's great capacity to decompose and recycles biodegradable pollutants in the domestic sewage. Materials such as glass, plastics, phenolics, metals, etc., and toxic substances such as mercury salts, pesticides, herbicides, cyanides, etc., that do not degrade (or degrade very slowly) under the natural environmental conditions are known as non-biodegradable pollutants. These pollutants are "biologically magnified", i.e., they pass from one biological system to another and get concentrated along the food chain.

6.4.2 Air Pollution

The earth's atmosphere is an envelope of gases extending to a height of about 2,000 km from the surface of the earth. The major gases in the atmosphere are nitrogen, oxygen, argon and carbon dioxide and traces of carbon monoxide, oxides of nitrogen and sulphur, hydrocarbons and particulates. The density of air decreases with increasing altitude and about 50% of the total mass of the atmosphere lie within the lowest 5 km (from the surface of the earth). The chemical composition of the atmosphere is modified by the injection into it of particles, gases and volatile substances, all are more or less toxic to living organisms. This hinders the renewal of certain natural components of the atmosphere in varying degrees and runs the risk of inducing irreversible ecological changes on a world-wide scale.

The World Health Organisation (WHO) has defined air pollution as "the presence of materials in the air in such concentrations which are harmful to man and his environment". In addition to industries, there are various other sources of air pollution in India, e.g., domestic combustion of low-grade fuels, fine dusts contributed by the deserts and other open and dry fields and human activities causing heavy pollution of the urban air (with suspended particulate matters and exhaust from motor vehicles). The main pollutants of air are carbon monoxide sulphur dioxide, oxides of nitrogen, carbon dioxide and particulates. Air pollution affects growth, living and life of humans, plants, organisms and all other living beings. It can be controlled by use of filter and electrostatic precipitator in chimneys, by installing catalytic converter in motor vehicles, using high-temperature incinerators and by developing non-pollution sources of energy.

6.4.3 Water Pollution

The hydrosphere includes water standing in lakes and swamps, flowing on the surface in rivers and streams, trapped within rocks below the surface of the earth, contained in the ocean basins and frozen in the polar ice caps. The water we use comes from two major sources, viz., surface water and ground water. Approximately 95% of the world's unfrozen fresh water exist as ground water. Of the total quantity of water in the hydrosphere, around 97% is in the oceans and only about 3% is fresh water. Of the total quantity of fresh water, 89% is not readily available. It has been estimated that, on an average, India gets about 60,000 billion litres of rain (including ice and snow) falls on the soil per day, out of which 40,000 billion liters of water return to atmosphere by evaporation and transpiration. The estimated commercial use of water in India at present is about 12,000 billion litres per day. Thus, reuse of surface water is the only way in India to satisfy the demand for water in the future.

Water pollution may be defined as "the addition of substances (organic, inorganic, biological or radiological) or factors (e.g., heat) which degrade the quality of water so that either it becomes a health hazard or unfit for use". The major sources of water pollution are :

(1) Domestic .
(2) Industrial wastes.
(3) Agricultural chemicals.

(4) Excess heat (from nuclear power plants, for example).

(5) Oil spills (from tankers or coastal oil wells).

Polluted water can be purified by filtration, chlorination, boiling and by treating waste water in water treatment plants. Recycling of urban, industrial and agricultural wastes can reduce water pollution to a great extent.

6.4.4 Soil Pollution

The principal causes of soil pollution are agricultural, urban and industrial wastes. Agricultural chemicals like insecticides, herbicides and fungicides, which are freely used by both developed and developing countries for obtaining larger agricultural yields, find their way into the soil. Since the degradation of these highly toxic compounds by biological processes is very slow, they persist in the soil for a long time and enter into the food chains beginning with plants and finally reaching humans. The urban wastes are mostly biodegradable and can be converted into valuable manure by composting. Industrial wastes, on the other hand, contain many substances which are toxic to the biological community. They cannot be easily decomposed by natural processes.

Many of the industrial solid wastes are dumped on the surrounding land and liquid wastes are discharged into streams through drains. The solid wastes directly pollute the soil. The liquid wastes are partly absorbed by the soil (thus polluting it) and the rest of it ultimately finds its way into the sea water, polluting the soil all along its way.

6.4.5 Noise Pollution

Noise may be defined as any sound unwanted by the listener. Noise is increasingly becoming a major pollutant of the environment. Noise has a subtle but definitely harmful effect on human health. High-level, jarring noise is a great nerve wrecker. It can result in headache, fatigue, dizziness, high blood pressure and abnormal heart rhythms. Prolonged exposure to high-level noise leads to noise-induced hearing loss.

Noise can be reduced by : (1) modifying the sources to reduce the noise output, (2) altering the transmission path (by using noise absorbers, for example), (3) enclosing the receiver or the source of noise or both, (4) regulating the exposure time and (5) providing personal protective devices (e.g., earplugs).

6.5 ECONOMIC DEVELOPMENT AND ENVIRONMENT

Although the desirability of economic development is universally recognized, recent years have witnessed an increasing concern as to whether environmental constraints will limit development and whether development will cause serious environmental damage, in turn, impairing the quality of life of the present and future generations. Some environmental problems are clearly associated with the lack of economic development. The obvious examples of such problems are inadequate supply of clean water, lack of sanitation, indoor air pollution from biomass burning and many types of land degradations in developing countries. In this case, the challenge is to accelerate equitable income growth and promote access to the necessary resources and technologies. On the other hand, industrial and energy-related pollution both local and global, deforestation caused by commercial logging and overuse of water resources are the results of economic growth that fail to take into account the importance of environmental protection. The challenge is to build the recognition of environmental security into the very process of decision making with or without development. Rapid growth of population may make it more difficult to address many environmental problems. We require, therefore, two types of policies, those that build on the positive links between development and the environment and those that break the negative links.

6.6 SUSTAINABLE DEVELOPMENT

The biggest contributing factor to the environmental degradation is the population explosion. The world's population is now growing at a rate of about 1.7% per year. About 90% of this growth take place in developing countries. For meeting the demands of the rising population, the natural resources are over-exploited and, if the current practices remain unchanged, this implies increased environmental damage. When the environment is damaged, human health is harmed, economic productivity is reduced and the pleasure obtained from the unspoiled environment is lost. The Report of the World Commission on Environment and Development has given a new direction in which the development efforts need to be directed for sustainability and environmental protection. The Report points out that the very content of economic growth has to be changed and that the consumption of material goods has to be ranked according to a scale of sustainability. Another important observation of this Report pertains to the holistic nature of the environment. According to the Report, the environment does not exist as a sphere separate from human actions, ambitions and needs. Any attempt to defend it in isolation from other human concerns gives the very word "environment" a connotation of naivety in some political circles. The Report has recognized the relationship between environmental and developmental concerns and has proposed a number of long-range steps for changing conditions and practices which affect the survival of humans. Many critical issues of survival are related to uneven development, poverty and population growth. These issues place unprecedented pressure on land, water, forests and other natural resources. The downward spiral of poverty and environmental degradation is a waste of opportunities and resources. What is needed now is a new ear of economic growth-a growth that is forceful and, at the same time, socially and environmentally sustainable.

6.7 THE ENVIRONMENTAL ETHIC

The word *"ethic"* is derived from the Greek work *"ethos"*, which means the character of a person as defined by his actions. This character has been developed during the evolutionary process and has been influenced by the need for adopting to the environment. The "ethic", in short governs our way of doing things and this is a direct result of our environment. Ecology and economics are on a collision course now. The conflict between the ecologists and economists has been aptly summarized by Kenneth Boulding :

> *"Ecology is uneconomic.*
> *But with another kind of logic, Economy is unecologic".*

In the ecological context, maladaptation to the environment by an organism results in two options :

(1) The organism dies out or
(2) The organism evolves to a form and character where it is once again compatible with the environment.

The acceptable option is one in which we must learn to share, in an equitable manner, our vast but finite natural resources to regain a balance. This requires that our needs be reduced and that the materials that we do use must be replenishable. The recognition of the need for such adaptation as a means of survival has developed into what we now call the *"environmental ethic"*. The birth of environmental ethic as a force is partly a result of our concern for our own long-term survival, as well as our realization that humans are but one form of life and that we should share our earth with our fellow travelers. Environmental ethic is not a religion since it is based not only on faith, but also on hard facts and through analysis. The environmental ethic is very new and none of the doctrine is cast in immutable decrees and dogmas. Education of the public to environmental problems and solutions is of prime importance.

6.8 THE ROLE OF ENVIRONMENTAL ENGINEER

As pollutants enter air, water or soil, natural processes such as dilution, biodegradation and chemical reactions convert waste materials to more acceptable forms and disperse them through a larger volume. However, these natural processes can no longer perform the clean-up alone due to the enormous amount of wastes generated now. The waste treatment facilities designed by the environmental engineer are mostly based on the principles of self-cleansing observed in nature, but the engineered processes amplify and optimize the operations observed in nature to handle larger volumes of pollutants and to treat them more rapidly. Environmental engineers adapt the principles of natural mechanisms to engineered systems for pollution control when : (1) they construct tall stacks to disperse and dilute air pollutants, (2) design biological treatment facilities for the removal of organic materials from waste water, (3) use chemicals to oxidize and precipitate iron and manganese in drinking water supplies, or (4) bury solid wastes in sanitary land-fill operations. Occasionally, the environmental engineer must also design to counteract or even reverse natural processes. For example, the containers used for the disposal of hazardous wastes (toxic chemicals or radioactive substances) must isolate those materials from the environment in order to prevent the onset of natural, but highly undesirable processes of dispersion and dilution.

An understanding of natural and engineered purification processes requires an understanding of the biological and chemical reactions involved in these processes. Thus, in addition to being knowledgeable in the mathematical, physical and engineering sciences, the environmental engineer must also be well grounded in the subject areas of chemistry and microbiology. In fact, an understanding of chemical and biological principles is as essential to the environmental engineer as the understanding of statics and strength of materials is to the structural engineer.

The unique role of the environmental engineer is to build a bridge between chemistry, biology and technology by applying all the techniques available by modern science and technology to the job of cleaning up the debris left in the wake of an indiscriminate use of technology. The laws of the conservation of mass and energy prevent the destruction of pollutants and the engineer is bound by these laws. Thus the objectives of waste treatment must be to convert the objectionable forms; to disperse the pollutants so that their concentrations are minimal, or to isolate them from the environment. In all instances, the end products of the treatment of polluted water or air (or of the disposal of solid wastes) must be compatible with the existing environmental resources and must not overtax the assimilative power of the hydrosphere, atmosphere or lithosphere. It is only by bringing technology into harmony with the natural environment that the environmental engineer can hope to achieve a major goal of his profession, viz., the protection of the environment from the potentially harmful effects of human activity, the protection of human population from the effects of adverse environmental factors and the improvement of environmental quality for human health and well-being.

MODEL QUESTIONS

(Essay/Long Type)

6.1 Give most suitable definitions of "environment" and "environmental pollution". What is a pollutant as defined by the Government of India in its Environmental Protection Act (1986)?

6.2 Write brief notes on atmosphere, hydrosphere and lithosphere.

6.3 State air pollution as defined by the W.H.O. List the major causes of air pollution.

6.4 What are the major sources of water pollution ? Explain in brief.

6.5 Explain in brief the environmental ethic.

6.6 Explain the role of an Environmental Engineer.

(Objective/Short Type)

6.7 List major four natural needs of mankind.

6.8 List major four acquired needs of mankind.

6.9 A pollutant may be :

    (a) Biodegradable   (b) Toxic

    (c) Only (b)   (d) Both.

6.10 Match the following :

    (A) Agricultural wastes   (a) Air pollution

    (B) Catalytic converter   (b) Motor vehicles

    (C) Incinerators   (c) Nonbiodegradable

    (D) Pesticides   (d) Soil pollution

6.11 Write true or false :

Domestic sewage has mainly biodegradable pollutants.

6.12 Name the odd ones :

    (a) Poverty   (b) Economic growth

    (c) Population   (d) Pollution.

6.13 Fill up the blanks :

The global environment consists of atmosphere, _____ and
_____ .

7
Chapter

Ecosystems

7.1 ECOLOGY AND ITS SCOPE

7.1.1 What is Ecology?

Life does not occur in a vacuum. Every living organism is surrounded by materials and forces which constitute its environment and the environment supplies all the needs of the living organism. For its survival, an animal or a plant requires from its environment a supply of materials (food, air and water) and energy (sunlight). For these basic requirements, each living organism has to depend upon and interact with the biotic (living) and abiotic (non-living) components of the environment. The biotic components of the environment comprise plants, animals and microbes, all of which interact in an energy-dependent system. The abiotic components of the environment, include basic inorganic substances such as oxygen, carbon dioxide, water, nitrates, phosphates, etc., and an array of organic compounds (the by-products of the organism's activities or death). They also include physical factors like soil, rainfall, temperature, moisture, winds, currents and solar radiation. The scientific study of the interactions of organisms with each other and with their physical environment is known as ecology. The main concerns of ecology are the directive influences of the biotic and abiotic environmental factors on the growth, distribution, behaviour and survival of organisms.

It was the German zoologist **Ernst Haeckel** who coined the term *"Oekologie"* in 1866 to refer to the inter-relationship of living organisms and their environment. In 1857, the American naturalist and philosopher **Henry David Thoreau** had also spoken of ecology among other fields of biology and natural history, but he did not provide a definition of this term. The word *"ecology"* comes from two Greek words, viz., *"Oikos"* (meaning "household", "place to live" or "habitation") and *"logos"* (meaning "study" or "discourse"). Thus, the science of ecology literally deals with the organism and its place of living. Basically, the organism's place of living is its environment; so ecology may also be called *"environmental biology"*.

7.1.2 Definitions of Ecology

The term *"ecology"* has been defined by different classical and modern ecologists with different views, and no universally accepted definition of ecology has been formulated by any ecologist so far. For the sake of better understanding of the scope, limitation, purpose and mode of study of different ecological phenomena, several definitions of the term *"ecology"*, given by various ecologists, are arranged below in a chronological order :

(1) Ernst Haeckel (1866) "Ecology is the body of knowledge concerning the economy of nature i.e., the investigation of the total relations of the animal to its inorganic and organic environment".

(2) Frederick Clements (1916) "Ecology is the science of community".

(3) Charles Elton (1927) "Ecology is the scientific natural history concerned with the sociology and economics of animals".

(4) W.P. Taylor (1936) "Ecology is the science of the relations of all organisms to all their environments".

(5) W.C. Allee (1949) "Ecology is the science of inter-relations between living organisms and their environment, including both the physical and biotic environments and emphasizing inter-species as well as intra-species relations".

(6) G.L. Clarke (1954) "Ecology is the study of inter-relations of plants and animals with their environment, which may include influences of other plants and animals present as well as those of the physical features".

(7) A.M. Woodbury (1955) "Ecology is the science which investigates organisms in relation to their environment – a philosophy in which the world of life is interpreted in terms of natural processes".

(8) A.Mac Fadyen (1957) "Ecology is a science which concerns itself with the inter-relationships of living organisms, plants and animals and their environments".

(9) S.C. Kendeigh (1961) "Ecology is the study of animals and plants in their relation to each other and to their environment".

(10) H.G. Andrewartha (1961) "Ecology is the scientific study of the distribution and abundance of organisms".

(11) L.R. Tayor (1967) "Ecology is the study of the way in which individual organisms, populations of some species and communities of populations respond to these changes".

(12) G.A. Petrides (1968) "Ecology is the study of environmental interactions which control the welfare of living things, regulating their distribution, abundance, production and evolution".

(13) Eugene Odum (1971) "Ecology is the study of the structure and function of ecosystems, or broadly the study of nature".

(14) C.J. Krebs (1972) "Ecology is the scientific study of the interactions that determine the distribution and abundance of organisms".

(15) M.E. Clark (1973) "Ecology is the study of ecosystems or the totality of the reciprocal interactions between living organisms and their physical surroundings".

(16) E. Pianaka (1973) "Ecology is the study of relations between organisms and the totality of biological and physical factors affecting them or influenced by them".

(17) C.H. SOUTHWICK (1976) "Ecology is the scientific study of the relationships of living organisms with each other and with their environments. It is the science of biological interactions among individuals, populations and communities".

(18) R.L. Smith (1977) "Ecology is a multidisciplinary science which deals with the organism and its place to live and which focuses on the ecosystem".

7.1.3 Scope of Ecology

By its very nature, ecology is a multidisciplinary science. It involves plant and animal biology, taxonomy, physiology, genetics, behavioural science, meteorology, pedology, geology, sociology, anthropology, physics, chemistry, mathematics and electronics. Often, it is very difficult to draw a sharp line between ecology and any of the above disciplines since all impinge on it. Exactly the same situation exists also within ecology. In order to understand the interactions between the organism and the environment, or between organisms, it is often difficult to separate behaviour from physiology, adaptation from evolution and genetics, or animal ecology from plant ecology.

Historically, ecology developed along two lines, viz., the study of plants (plant ecology) and the study of animals (animal ecology). Plant ecology focuses on the relationship of plants to other plants and their environment. The approach of plant ecology is largely descriptive of the vegetational and floristic composition of an area and usually ignores the influence of animals on the plants. Animal ecology, deals with the study of population dynamics, population distribution, animal behaviour and the relationships of animals and their environment. Because animals depend upon plants for food and shelter, animal ecology cannot be fully understood without a considerable background of plant ecology. This is particularly true in the areas of applied ecology like wildlife management.

Both plant and animal ecology may be approached as the study of the interrelations of an individual organism with its environment (*autecology*), or as the study of groups of organisms (*synecology*). In many ways, autecology is the classical study of ecology, which is experimental and inductive. Because autecology is usually concerned with the relationship of an organism to one or more variables like humidity, light, salinity or nutrient levels, it is easily quantified and lends itself to experimental design both in the laboratory and in the field. Autecology has, therefore, borrowed experimental techniques from physics, chemistry and physiology.

Synecology, is philosophical and deductive in nature. It is largely descriptive and not easily quantified. Some of the important concepts developed by synecology are those concerned with nutrient cycles, energy budgets and ecosystem developments. Synecology has strong ties with geology, meteorology and cultural anthropology. Synecology may be subdivided according to environmental types, e.g., terrestrial or aquatic ecology. Terrestrial ecology may be further subdivided into forest, grassland, desert and arctic ecology. It concerns with such aspects of terrestrial ecosystems as microclimate, soil chemistry, soil fauna, hydrologic cycles, ecogenetics and productivity. Terrestrial ecosystems are more influenced by organism and are subject to much wider environmental fluctuations than aquatic ecosystems. Because the physical environment is so important in controlling the aquatic ecosystem, considerable attention is paid to the chemical and physical characteristics of the ecosystem, such as the current and the chemical composition of the water. By convention, aquatic ecology (known as limnology) is limited to stream ecology and lake ecology. Stream ecology concerns life in flowing waters, while lake ecology deals with life in relatively still water. Marine ecology, on the other hand, concerns life in open seas and estuaries.

Other ecological approaches concern specialized areas. The study of the geographic distribution of plants and animals is known as ecological plant and animal geography. The study of population growth and mortality, is known as population ecology. The study of the genetics and ecology of local races and distinct species is called ecological genetics. The study of the behavioural responses of animals to their environment is known as behavioural ecology. Investigations of the interactions between the physical environment and the organism belong to ecoclimatology and physiological ecology. The study of the groups of organisms is known as community ecology. The part of ecology concerned with the analysis and understanding of the structure and function of ecosystems by using applied mathematics, mathematical models and computer programmes is known as systems ecology. Applied ecology, on the other hand, is concerned with the applications of ecological principles to the management of natural resources, agricultural production and problems of environmental pollution.

7.1.4 Objectives of Ecology

Ecology is a distinct science because of the following reasons :
  (1) Ecology is a body of knowledge not similarly organized in any other branch of biology.
  (2) Ecology uses a special set of techniques and procedures.
  (3) Ecology has a unique point of view.

The essence of ecology lies in a comprehensive understanding of the following phenomena :
  (1) The local and geographical distribution and abundance of organisms (study of habitat, niche, community and biogeography).
  (2) Temporal changes (seasonal, annual, successional and geological) in the occurrence, abundance and activities of organisms.
  (3) The inter-relations between organisms in populations and communities (population ecology).
  (4) The structural adaptations and functional adjustments of organisms to their physical environment (physiological ecology).
  (5) The behaviour of organisms under natural conditions (ethology).
  (6) The evolutionary development of all the above interrelations (evolutionary ecology).
  (7) The biological productivity of nature and the way it serves mankind.
  (8) The development of mathematical models to relate interactions of various parameters and prediction of the effects of these interactions (ecosystems analysis).
  (9) The conservation and management of natural resources and control of environmental pollution (applied ecology).

7.1.5 Experimental Techniques

Because ecologists work with living systems possessing numerous variables, the techniques used by physicists, chemists, engineers and mathematicians require modifications before they can be used in ecological investigations. The experimental results in ecology are not as precise as those obtained in other sciences. It is relatively simple for a physicist, for example, to measure gain or loss of heat from metals (or other animate objects), which possess certain physical properties such as thermal conductivity, expansion coefficient, surface features, etc. To determine the heat exchange between an animal and its environment, on the other hand, a physiological ecologist is confronted with an array of almost unquantifiable variables. He, moreover, has the formidable task of gathering numerous data and then analyzing them. In spite of these limitations, various aspects of the environment can be determined by physical and chemical techniques, ranging from simple chemical identification and physical measurement to the use of sophisticated electronic apparatus. Development in biostatistics, experimental designs and sampling techniques permit a quantitative statistical approach to the study of ecology. Because of the extreme difficulties of controlling environmental variables in the field, investigations involving the use of experimental design are largely confined to the laboratory and to the controlled field experiments designed to test the effects of only one or just a few variables. The use of statistical techniques and the application of computer science to mathematical models based on data obtained from the field are providing new insights into population dynamics and functioning of the ecosystems. Mathematical programming is gaining an increasing importance in applied ecology, especially in the management of natural resources and agricultural problems.

Controlled environmental chambers enable experimenters to maintain plants and animals under known conditions of light, temperature, humidity and time so that the effects of each variable (or a combination of variables) can be studied. Biotelemetry and other electronic tracking devices permit the rapid and nondestructive sampling of plant and animal populations. Such tools enable ecologists to follow from a distance the movements and behaviour of a free-ranging animal by radio signals beamed from a small transmitter attached to the organism. Radioisotopes are used for tracing the pathways of nutrients through ecosystems for determining the time and extent of transfer of energy and nutrients through different components of the ecosystem and also for the determination of food chains. The use of laboratory micro-ecosystems, consisting of biotic and nonbiotic materials from natural ecosystems, held under conditions similar to those found in the field, are useful in determining ratio of nutrient cycling, ecosystem development and other functional aspects of ecosystems.

7.1.6 Applications of Ecology

Man, the dominant organism on earth, is discovering that he is closely tied to his environment. The air he breathes, the water he drinks, the food he consumes and the products he uses and throws away bind him to the local and global ecosystems. Agriculture, forestry, wildlife and fisheries-all represent some aspects of man's modification and exploitation of natural ecosystems. In order to achieve maximum food production, man has eliminated stable and complex ecosystems and replaced them with fragile and simplified single-species ecosystems (or monocultures), in which the energy flow is directed to crops. He has shortened food chains by selecting and breeding domestic live stock that produces milk, meat or wool by converting grasses directly into products useful to man. Successful agriculture depends upon the ecological adaptation of crop plants and domestic animals to environmental conditions and upon the genetic capability of plants to respond to agricultural practices like fertilization. Amount of fertilizers and nutrients entering into farm ecosystems in some instances, has undesirable side-effects. Excess phosphates and nitrates stimulate algal blooms (excessive growth of algae) in ponds and lakes which deplete the oxygen dissolved in water and suffocate fish and other aquatic organisms. Simplified ecosystems also attract pest species (both plant and animal), requiring costly control methods, which often have adverse environmental effects. Overgrazing of grasslands and mismanagement of crop fields result in the erosion of soil by wind and water. The result is that topsoil is carried to rivers and lakes and agricultural lands are impoverished. Management of forests involves the manipulation of natural forests by means of cutting and controlled burning to maintain the forest in a stage of development that favours those species which are most desirable economically. Management and exploitation of the fish populations for sustained yields require a thorough knowledge of fish biology and of the essential physical and biological aspects of the environments. Failure to give careful consideration to these points (especially to the dynamics of populations) threatens to extinguish certain oceanic fish species, thus destroying the fishing industry and removing an important source of protein for man.

Environmental pollution represents a short-circuiting of the biosphere cycles (i.e., the chemical or nutrient cycles) in nature. More materials are being poured into the global ecosystem (or biosphere) than can be recycled naturally. As these materials accumulate in the biosphere, they have adverse, and often lethal effects on various organisms. Too much sulphur dioxide in the atmosphere can kill and injure plants and animals. Unnatural substances also enter the biosphere cycles. Chlorinated pesticides (like DDT and BHC) are circulated by the atmospheric and oceanic currents to all parts of the earth, where they are taken up and concentrated in several tropic levels. In large doses, these pesticides can threaten the survival of organisms like birds by impairing their metabolism and reproduction. Radioactive contaminants often enter metabolic cycles, thus presenting a serious health hazard. Other man-induced changes in the ecosystem can create equally severe public health problems. In certain tropical regions, the replacement of simple age-old irrigation practices with large-scale dam projects and irrigation channels (filled with warm and slow-moving water) has created ideal living and breeding conditions for mosquitoes and snails that serve as secondary hosts for organisms that cause parasitic diseases in humans and animals. Atmospheric pollutants, on the other hand, are responsible for a distinct rise in cases of chronic bronchitis, nasal and eye irritations and lung cancer. The health and physical well-being of man and even his continued survival depends on how he applies the principles of ecology towards the solution of many urgent environmental problems.

7.2 NATURAL ECOSYSTEMS

7.2.1 Concept of the Ecosystems

For a long time, the science of ecology lacked a strong conceptual base. Modern ecology, however, is now focused on the concept of the *eco-system*, which is a functional unit consisting of interacting organisms and all aspects of the environment in any specific area. It contains both the nonliving (abiotic) and living (biotic) components through which nutrients are cycled and energy flows. To accomplish this nutrient cycling and energy flow, ecosystems must posses a number of structured interrelationships between soil, water and nutrients on the one hand and producers, consumers and decomposers on the other hand.

Ecosystems function by maintaining a flow of energy and a cycling of nutrients through a series of steps of eating and being eaten, of utilization and conversion, which is known as *food chain.* Ecosystems tend towards *maturity* or stability and, in doing so, they pass from a less complex to a more complex state. This directional change is known as *succession.* The major functional unit of the ecosystem is the *population.* It occupies a certain functional niche, related to its role in nutrient cycling and energy flow. Both the environment and the amount of energy fixation in any given ecosystem are limited. When a population reaches the limits imposed by the ecosystem, its numbers either stabilize or decline from disease, starvation, strife, low reproduction, or other behavioural or physiological reactions. Changes and fluctuations in the environment represent selective pressures upon the population to which it must adjust. The ecosystem has historical aspects since the present is related to the past and the future is related to the present. Thus the ecosystem is the one concept that unifies plant and animal ecology, population dynamics, behaviour and evolution.

7.2.2 Basic Features of the Ecosystems

There is an interdependence between the plants and animals of a biotic community and an interchange of materials takes place between them. A tree, for example, provides shelter to the birds and some other animals and supplies them with food and oxygen. The birds and animals, on the other hand, supply carbon dioxide required by the tree for photosynthesis. Dead plants and animals and animal excretions serve as food for the soil fungi and bacteria, who convert these materials to organic manure, which is essential for plant life. The biological community and the non-living environment thus interact with each other and an exchange of materials takes place between them. Living organisms and their non-living environment are interdependent and together, they form an integrated unit in which the functions of its various components are well coordinated for the well-being of the entire unit. Such a natural unit of living community and non-living environment is known as an *ecological system* or simply an *ecosystem,* which is the basic functional unit of ecology. The study of ecology is, in fact, the study of ecosystems.

The scientific term *"ecosystem"* is the equivalent of *"nature"* in common language, as understood by a layman. The term *"ecosystem"* was first used by the British scientist A. G. Tansley in 1935. An ecosystem may be as small as a small quantity of water in a dish, or a small lump of soil or a single tree. An ecosystem can also be as large as a forest, or an ocean or even the whole earth. Quite often several ecosystems are interlinked and interact with one another. Small ecosystems form a part of larger ecosystems. A bird, with several populations of parasites living on and inside its body, constitutes a small ecosystem. A single tree, providing food and shelter to a large number of birds and other animals, is comparatively a larger ecosystem. The individual tree, on the other hand, is part of a still larger ecosystem, viz., the forest, which contains many other kinds of trees, shrubs, herbs and a wide range of animals.

There are some basic features which are common to all ecosystems small or large, on land or in water. The biotic community in an ecosystem requires energy for its maintenance and growth. With the exception of a small group of chemosynthetic bacteria which obtain their energy directly from chemical reactions, the ultimate source of energy for all other plants and animals is the sun. Only the chlorophyll-containing plants are capable of capturing the energy of sunlight and converting it into the chemical energy of organic compounds by the process of photosynthesis. In this process, the plants obtain carbon dioxide from air, or that dissolved in water (in the case of aquatic flora) and combine it with water to produce organic food like carbohydrates, fats and proteins and a large number of complex organic compounds, using the energy of solar radiation. All non-green plants like fungi and bacteria and all the animals in the ecosystem including man obtain their food requirements (i.e., energy) directly or indirectly from the green plants. When we take food in the form of wheat, rice, pulses and other seeds, vegetables, fruits, milk and meat, we are consuming the products of plant photosynthesis for the building of our body and for our energy requirements.

The green plants and through them the animals in the ecosystem, also obtain from the environment carbon, nitrogen, phosphorus, calcium, potassium, iron and other mineral nutrients and water. All these materials are ultimately returned to the environment for reutilization; so the material nutrients pass from the environment to the living organisms and then back to the environment. Thus, a continuous exchange of materials takes place between the environment and the living organisms. In case of energy, the flow occurs only in one direction, i.e., from a higher to a lower level (from the sun to green plants and then to animals). The *one-way flow of energy* and the *cycling of nutrients* are two important features of the ecosystem. They are responsible for the maintenance and regulation of the ecosystem.

The non-living environment is an essential part of the ecosystem since it supplies the nutrients and energy in the form of sunlight, in addition to controlling the functioning of the ecosystem and the interrelationships among the members of its biotic community. The biotic community also interacts with the environment and even modifies it; but ultimately it is the environment that determines and sustains the structure of the ecosystem. The driving force of the ecosystem is the energy of sunlight, which sustains the ecosystem. Based on the two basic features discussed above, an ecosystem may be defined as *an energy-based system which embraces the living organisms and non-living environment, both constantly influencing the functioning of each other and together making possible the continued development of life.*

Another important feature of an ecosystem is the property to resist change in its biological community and environment. Slight fluctuations may occasionally occur but, on the whole, the ecosystem tends to remain stable. The tendency of the ecosystem to maintain a relatively stable state of equilibrium between its different (but interdependent) parts is known as *homeostatic.* This is achieved by self-regulating mechanisms of *checks and balances* or forces and counter-forces which operate in all parts of the ecosystem and keep the fluctuations under control. For example, the biological community consumes large amounts of carbon dioxide and oxygen, yet the amount of these gases in the environment remains almost constant. This is because the consumption of oxygen and production of carbon dioxide in respiration by animals are counterbalanced by the opposing process of photosynthesis by plants, in which carbon dioxide is consumed and oxygen is released into the atmosphere. Any large disturbance in one component of the ecosystem affects the other components and upsets the balance in the ecosystem. This may lead to the development of a new structural and functional balance. As a result, the existing organisms may perish or be replaced by new ones, which may fit in to the changed environment. Man is now becoming conscious of the consequences of disturbing the balance in nature and of the need to control the ecosystems for his own survival.

7.2.3 Components of Ecosystem

Ecosystem consists of two major components, the living community (biotic) and the non-living environment (abiotic). From the point of view of *nutrition*, the biotic community is divided into two functional components:

(1) *Autotrophs* (self-nourishing).
(2) *Heterotrophs* (other-nourishing).

Autotrophs are chiefly the green plants which can produce their own organic food from simple inorganic substances like water, carbon dioxide and mineral elements, in the presence of sunlight, by using the process of photosynthesis. In this process, the energy of solar radiation is transformed into the potential energy of organic food molecules such as carbohydrates, proteins and fats. *Heterotrophs* are the fungi, the bacteria and the animals. Since these organisms have no chlorophyll to synthesise their food by photosynthesis, they obtain their food directly or indirectly from the autotrophs.

From the *structural point of view*, the ecosystem is made up of four constituents :

(1) Abiotic substances.
(2) Producers.
(3) Macroconsumers.
(4) Microconsumers or decomposers.

Abiotic substances in the ecosystem are water, oxygen, nitrogen, carbon dioxide, nutrient elements and compounds of calcium, potassium, iron, phosphates and nitrates. *Producers* are the autotrophs, mainly the green plants, which trap the thermal and optical energy of sunlight and produce organic food by photochemical reactions (photosynthesis). This food is necessary for the maintenance of life in the ecosystem. *Macroconsumers* are the large heterotrophic organisms, chiefly animals including man, that ingest other organisms or particulate organic matters. *Micorconsumers* or *decomposers* are small heterotrophs, mainly the bacteria and fungi, which decompose the dead remains of plants and animals, utilize a part of the products of this decomposition for their own nourishment and, in the process, release simple substances which can be reutilized as nutrients by the producers.

The well-known ecologist E.P. Odum has divided the components of the ecosystem into six classes. The first three of these comprise the *abiotic* components, while the last three make up the *biotic* components. These six classes of the components of the ecosystem are :

(1) Inorganic substances.
(2) Organic compounds.
(3) Climate regime.
(4) Producers.
(5) Macroconsumers.
(6) Microconsumers.

Inorganic substances in the ecosystem are water, chemical elements like iron, calcium, sodium, potassium, magnesium and phosphorus, and gases like oxygen and carbon dioxide. These constitute the nutrients or raw materials for green plants. *Organic compounds* in the ecosystem are the proteins, lipids, carbohydrates, fats, etc., contained in the dead plants and animals and the intermediates and end products of their decomposition (like urea and humus). The plant and animal contain many minerals and salts in bound form. As a result of the activity of microconsumer bacteria and fungi, the various elements bound in minerals are converted into inorganic form and are reutilized by the green plants. The organic matter links, therefore, the biotic and abiotic components of the ecosystem. The *climate regime* is the physical part of the environment. It includes factors like light, temperature and wind. The principal physical component is the solar energy, which is trapped by the green plants in photosynthesis and stored in the form of chemical energy of organic molecules. It is this energy that flows through the entire biotic community and makes life possible on the earth. Producers, Macroconsumers and Microconsumers have been defined earlier.

7.2.4 Types of Ecosystems

Ecosystems may be divided into two major classes :

(1) Natural ecosystems.
(2) Man-engineered ecosystems.

The natural ecosystems operate under natural conditions, while the man-engineered ecosystems are artificial ecosystems, which are maintained by man through planned manipulation.

The natural ecosystems may be further divided into two major classes :

(1) Terrestrial ecosystems.
(2) Aquatic ecosystems.

The terrestrial ecosystems include grassland ecosystem, forest ecosystem, desert ecosystem, mountain ecosystem, etc. Aquatic ecosystems, on the other hand, include two sub-classes :

(1) Fresh water ecosystems.
(2) Marine ecosystems.

The freshwater ecosystems include pond ecosystem, lake ecosystem, stream ecosystem, river ecosystem, etc. Marine ecosystems, include estuary ecosystem, sea ecosystem, seashore ecosystem, etc. In man-engineered ecosystems, man plans and keeps up the natural balance between biotic and abiotic components. Examples of such man-engineered ecosystems are: (1) cropland ecosystem such as a field of rice, wheat or maize and (2) microecosystems (those made in the laboratory for a planned study).

Some of the important differences between the terrestrial and aquatic ecosystems are as follows :

(1) The Size of the Producers

The most important difference between the terrestrial and aquatic ecosystems is the size of producers, i.e., green plants. The producers in terrestrial ecosystems are smaller in number, but bigger in size. This difference becomes quite impressive if we compare the number and size of planktons in a lake with the number and size of trees in a forest.

(2) Supporting Tissues

Land plants (i.e., the producers in terrestrial ecosystems) invest a larger part of their productive energy in their supporting tissues than those in aquatic ecosystems. The supporting tissue has a high content of cellulose and lignin, so it needs very little energy for its maintenance since it is resistant to consumers.

(3) Turnover

The phytoplanktons (tiny green plants) in a pond may replace themselves in a day when the pond metabolism is at its peak, but the plants in a terrestrial ecosystem live much longer, so the rate of their turnover is much less. The concept of turnover is particularly useful in dealing with the exchange of nutrients between organisms and the environment of an ecosystem.

(4) Amount of Detritus

In case of terrestrial ecosystems, there is a large amount of fibrous detritus (dead parts of plants and animals such as dry leaves, dead wood, etc.) because of the large structural mass of land plants and animals. In the aquatic ecosystems, by contrast, the detritus comes mainly from small organisms like planktons, so its amount is relatively smaller compared to that on land.

(5) Need for Water

In terrestrial ecosystems the amount of water needed by the producers (green plants) is much more than in aquatic ecosystems. In a grassland or forest ecosystem, for every gramme of carbon dioxide fixed by the process of photosynthesis, about 100 grammes of water is moved from the soil through the plant tissues (transpired and evaporated from the surfaces of plant leaves). No such massive amount of water is required by phytoplanktons (microscopic green plants) in an aquatic ecosystem.

7.2.5 Pond as an Ecosystem

A pond is a good example of a small, fresh-water, aquatic ecosystem, which is self-sufficient and self-regulating. In fact, one of the best ways to start the study of ecology is to study a small pond, where all the basic components of an ecosystem can be conveniently examined.

Abiotic Components

The abiotic components of a pond ecosystem consist of the physical and chemical parameters of pond water such as colour, odour, taste, turbidity, electrical conductivity, temperature, pH (hydrogen ion concentration), suspended solids, dissolved solids, dissolved oxygen, alkalinity, chemical oxygen demand (COD), biochemical oxygen demand (BOD), nitrate, sulphate, phosphate, chloride, etc., and certain minerals like sodium, potassium, calcium, magnesium, iron, manganese, nickel, cobalt, lead, mercury and arsenic. Water samples can be collected from the pond and all the above-mentioned physico-chemical parameters can be determined in a well-equipped laboratory.

Biotic Components

In a pond, the auto-trophic (self-nourishing) green plants and some species of photosynthetic bacteria are the *producers*. The green plants are mainly the *macrophytes* and *phytoplanktons* present in the pond water. The macrophytes are the rooted larger plants, which may be partly or completely submerged, emergent or free-floating. The phytoplanktons, on the other hand, are extremely small, floating or suspended lower plants. Diatoms form a major part of phytoplanktons. Other types of phytoplanktons are the green algae such as oscillatoria, volvox, spirogyra, etc.

In a pond, the majority of *consumers* are the herbivores. There are also a few insects and fishes which are carnivores. They depend on herbivores for their food. The *primary consumers* are either benthos or zooplanktons (microscopic animals). Benthos are the organisms present at the bottom of the pond, but some are associated with the living plants. The common examples of benthos are fishes, beetles or insect larvae. Zooplanktons are mainly rotifers, protozoans and crustaceans. They feed on phytoplanktons. Some fishes, insects and frogs come under the category of *secondary consumers*. They feed on zooplanktons or on small fishes. The large fishes that eat small fishes and water snakes that feed on fishes come under the category of *tertiary consumers*.

In a pond, there are some microorganisms that decompose the dead and decaying bodies of both consumers and producers. They are known as *decomposers*. Some bacteris, fungi and actinomycetes belong to this category of the biotic components of a pond. Their main function is to bring about the decomposition of dead and decaying organisms of the pond ecosystem. Thus, within the small area of a pond, all the important features of an ecosystem can be observed.

7.3 PRINCIPLES OF ECOBALANCE

7.3.1 Homeostasis in the Ecosystem or Principles of Ecobalance

The influence of animals on the ecosystem is mostly destructive. An enormous increase in the number of animals at any trophic level (a stage in a food chain) leads to a large-scale destruction of the preceding trophic level. Similarly, large-scale destruction of a higher trophic level will lead to a great increase in the population of the preceding trophic level. In both cases, the result is that the balance (or equilibrium) in the ecosystem is upset. This, however, is prevented by the fact that, while a given trophic level is a *consumer* with respect to the preceding level, it acts as a *secondary producer* for (and is, in turn, consumed by) the next higher trophic level. This system of checks and balances is an important stabilizing factor in any ecosystem. It acts as a self-regulating mechanism (i.e., a negative feedback loop), which prevents any large change in the living components of the ecosystem and maintains it in a state of equilibrium. The operation of this mechanism in nature is known as *homeostasis*. It may also be called the *principles of ecobalance.*

The ecological balance or ecobalance is generally maintained in natural ecosystems; but man's activities to exploit an ecosystem may disturb the balance, as illustrated by a simple example. India is a large country with more than 70% of its population engaged in agriculture and agro-industries. The agriculture produce is a matter of vital concern to the nation. A common herbivore of our food grain crops is the rat. In nature, wild cats, snakes, kites, hawks and owls feed upon the rat and keep its population under control. However, man's interference with the working of nature has caused great diminution in the population of these predators. As a result, the population of rats has greatly increased and this, in turn, has caused great damage to our grain and other crops. The loss suffered by the nation on this single count is enormous. It has been estimated that approximately 20% of the food grains produced in India are destroyed by rats and other pests. This huge loss can be reduced only by application of the principles of ecological balance and biological control. It is very difficult to control the rat population unless measures are taken to protect and perhaps increase the present population of the natural consumers of rats such as snakes, cats, hawks, etc. Man disturbs the natural controls based on the principle of ecobalance (or homeostasis) and, in their place, introduces artificial controls (such as pesticides and herbicides) for his own benefit. The natural ecosystems are manipulated and run by artificial methods to enable mankind to obtain optimum production from the ecosystem. This artificial manipulation of natural ecosystems (e.g., artificial irrigation or the use of chemical fertilizer) disturbs the natural balance of the ecosystem, which may have highly adverse consequences in the long term.

7.3.2 Balance of Nature

All the components of an ecosystem are very closely interrelated for its smooth functioning. This close interrelationship between the components of an ecosystem is based on food chains, cycling of nutrients and other materials and energy flow. Each individual organism depends upon other organisms for food and other requirements for survival and perpetuation of the species. Since the organisms depend upon one another, there must be a balance in their number; and when the population of one species is disturbed by a large increase or decrease, the population and activities of other interrelated species are affected. Consider, for example, the number of a herbivore like a deer and the number of a carnivore like a tiger in a forest. If the deer population is decreased due to any reason, it will adversely affect the tiger population. On the other hand, if grasses decrease on the grassland, that will adversely affect the deer population since grasses form the food for the deer population. In the absence of tiger population, however, the deer population may increase rapidly and deplete the grasses in a relatively short period and the dear population may start migrating to other areas. Thus the grass, deer and tiger are dependent upon one another for their existence, continuance and perpetuation. This is an example of the *"Balance of Nature"* or *"Ecobalance"*.

In all groups of plants and animals that live side by side in an ecosystem, there have to be some who hunt and some who are hunted. There must be some animals who eat plants and some who feed on other animals. This always leads to the phenomenon of ecobalance. If all the animals were herbivores, there would soon be too many of them. They would denude the ecosystem by eating away all the plants in a short time and then they would perish, in turn, due to lack of food. By having some animals in the ecosystem that feed on other animals, their number is kept down and there is a balance between various kinds of animals and their food. In this way, there is a chance for many (both plants and animals) to coexist and thrive on the earth.

For the most part, all animals develop according to their eating habit. There is always an alternate source of food for a particular animal so that the stability of the ecosystem is maintained. The dogs eat rabbits, and the rabbits eat herbs. However, if dogs eat only rabbits, the number of rabbits will decrease and in a short period, the rabbits will not be available any more. In the absence of rabbits, however, dogs will not die since they will eat mice. When the number of rabbits decreases, herbs will have a better chance to grow. As a result, mice will have more food and better place for hiding, so their number will increase. The dogs, in turn, will have sufficient mice to eat and this will lead to a reduction in the number of mice. Moreover, rabbits will increase their population when they are less hunted by dogs (who depend more on mice when less rabbits are available). In this way, a balance is always maintained in a natural ecosystem. As long as the environment does not change, there is no large change in the ecosystem. It is man who upsets the ecobalance by various activities such as deforestation, large irrigation projects, mining, industrialization, urbanization, etc. In nature, every plant and every animal has a certain role to play. In the name of development, man should not destroy ecological balance to face drastic consequences in the years to come.

7.4 BIOSPHERE CYCLE

7.4.1 Biosphere or Biogeochemical Cycle

The total mass of all the organisms that have lived on our planet in the past 1.5 billion years, the period during which life has existed on earth, is much greater than the mass of carbon, nitrogen, oxygen and hydrogen atoms, the major constituents of the bodies of plants and animals present on the earth. According to the *law of conservation of matter,* the matter can neither be created nor destroyed. Obviously, carbon, nitrogen, etc., must have been used over and over again in the bodies of plants and animals that have existed on earth over the course of time. The earth neither receives any great amount of matter from the outer space nor does it lose significant amount of matter to it. The atoms and molecules of elements such as carbon, hydrogen, oxygen, nitrogen, phosphorus, calcium, magnesium, etc and the rest found in plant and animal tissues are taken from the environment by a circuitous route involving several other organisms and are returned to the environment to be used again. Such cyclic movements of chemical elements of the biosphere between organisms and the environment are known as the *biogeochemical cycles* or simply *biosphere cycles.*

All living organisms require various chemical elements for their metabolic process and growth. Some elements like carbon, hydrogen, oxygen, nitrogen and phosphorus form the basic constituents of the body and are essential elements for the protoplasm. Elements like sulphur, magnesium, calcium, potassium and iron help in the metabolic process. Carbohydrates and fats are formed by carbon, hydrogen and oxygen. Protein is synthesized with nitrogen and the above three elements. Nucleic acids are formed by carbon, hydrogen, oxygen, nitrogen and phosphorus. Magnesium, in the other hand, is essential for the formation of chlorophylls in plants, whereas sulphur is the main element in the synthesis of aminoacids. Thus, the various chemical elements play an important role in the survival and growth of organisms.

Some elements like carbon, hydrogen, oxygen, nitrogen, phosphorus, calcium, magnesium, sulphur and potassium are required in relatively large quantities; so they are called *macronutrients*. Some other elements like iron, manganese, copper, zinc, boron, molybdenum, cobalt, sodium and chlorine are required in relatively small amounts and they are known as *micronutrients*. All these elements maintain their continuity by circulating in characteristic paths or cycles from the environment to organisms and back to the environment. Living organisms play significant role for maintaining these cycles in the biosphere. Without the organisms, biosphere cycles will not be possible.

Biosphere cycles can be classified into three groups :

(1) Hydrological cycle or water cycle.
(2) Gaseous cycles such as oxygen, nitrogen and carbon dioxide cycles.
(3) Sedimentary cycles such as sulphur and phosphorus cycles.

Some important biosphere cycles are described below :

7.4.2 Carbon Dioxide Cycle

7.4.2.1 Basic Ideas

The bulk of living matter, plants and animals, is made up of four chemical elements; carbon, hydrogen, oxygen and nitrogen. For life to continue on earth, these four critical elements must move in interrelated cycles throughout the biosphere. The carbon cycle is generally termed as the **carbon dioxide cycle**, which is perhaps the simplest of the nutrient cycles in the biosphere. This cycle is essentially a perfect one in the sense that carbon dioxide (CO_2) is returned to the environment about as fast as it is removed by the processes of photosynthesis.

7.4.2.2 The CO_2 Cycle

Since carbon is a basic constituent of all organic compounds and a major element involved in the fixation of energy by photosynthesis, the carbon dioxide cycle is so closely tied to the energy flow in an ecosystem that the two are inseparable. During photosynthesis, atmospheric carbon dioxide is incorporated into the production of the carbohydrate known as **glucose** ($C_6H_{12}O_6$), which is subsequently converted to more complex organic compounds like polysaccharides (sucrose, starch, cellulose), proteins and lipids. All the polymeric organic compounds containing carbon are stored in different plant tissues as food. From the plants, carbon is passed on to the trophic levels of herbivores or retained by the plant until it dies and serves as food for the decomposers. A part of this carbon is returned to the atmosphere or the enveloping aqueous medium. The carbon dioxide, which is released as the byproduct of plant and animal respiration, is used again by plants in photosynthesis. Decomposing microorganisms are important in breaking down the dead material with the release of carbon back into the carbon cycle. Similarly, the carbon taken up by herbivores may travel along a number of routes. It may be incorporated into the protoplasm and stored until the organism dies, whereupon it is utilised by the decomposers. Alternately, the carbon may be released to the atmosphere as CO_2 through animal respiration or it may serve as live food for other animals (the carnivores). A similar fate awaits the carbon taken up by the carnivores (animals feeding on other animals).

The carbon compounds that are lost to the food chain after fermentation, such as methane (CH_4), are readily oxidized to carbon dioxide in the atmosphere by inorganic reactions. A part of the carbon stored in sediments may be released by erosion. Chemical weathering of rocks may oxidise some of the carbon contained in the rocks and release the resulting CO_2 into the atmosphere. Carbon permanently trapped in the rocks and not uncovered by weathering may be released as carbon dioxide from volcanoes. During the last one hundred years or so, man has greatly increased the rate at which carbon is passing from its sedimentary form to the atmospheric carbon dioxide by burning the fossil fuels like oil and coal. The combustion of fossil fuels is a significant means of recycling sedimentary carbon much faster than natural weathering or volcanic eruptions.

A small portion of carbon especially in the sea is found not as organically fixed carbon, but as carbonates, especially as calcium carbonate ($CaCO_3$), which is commonly used for shell construction by animals like oysters. The carbon dioxide dissolved in water reacts with it to form carbonate (CO_3). Aquatic organisms like oysters can combine the carbonate with the calcium dissolved in water to produce calcium carbonate. After death of these animals, the calcium carbonate may either dissolve in water or remain in sedimentary form.

The carbon dioxide (or carbon) cycle may be summarized as follows:

(1) Carbon undergoes a complete cycle as it moves from the atmospheric carbon dioxide and that dissolved in water to terrestrial and aquatic

plants through photosynthesis and from the plants through the food chains to the herbivores and the carnivores.

(2) At each step, a part of the carbon stored in complex food molecules is broken down to provide energy to the organism and returned to the air and water as carbon dioxide by the process of cellular respiration.

(3) The remaining carbon is also returned to the air and water, when the plant or the animal eventually dies and is broken down into carbon dioxide by the bacteria and fungi that decompose organic matters.

(4) When the atmospheric and aquatic carbon dioxide is used by plants in the process of photosynthesis, the solar energy used to drive the carbon dioxide cycle is converted first to chemical energy in the chemical bonds of larger and complex organic molecules. The heat energy thus released flows back into the outer space.

(5) Upon weathering of carbonate rocks, combustion of fossil fuels and volcanic activity, the carbon bound in sedimentary rocks is returned to the atmospheric or oceanic reservoir.

(6) The most important feature of this cycle is that carbon is utilised by a number of avenues and then restored to the atmosphere. Collectively, these various avenues (or pathways) constitute self-regulating feedback mechanisms resulting in a relatively homeostatic or stable system.

7.4.2.3 Special Features

There are certain control mechanisms which are inherent in the carbon dioxide cycle. The rate of carbon utilization is obviously dependent on its availability. If excessive amounts of carbon are taken up in any one phase of the cycle, other phases of activity may be inhibited or slowed down. For example, if the pH of water is in the alkaline range, more carbon is tied up as carbonate and less of it is in the solution form. This removal of carbon from solution would upset the equilibrium established between the atmospheric and dissolved carbon dioxide. That would result in a movement of CO_2 from the atmosphere to solution until the equilibrium is reached.

Though the carbon dioxide cycle exhibits a basic similarity with other biosphere cycles, it has a peculiar feature that its organic phase is not essentially a complete cycle within itself. The organic or biotic and atmospheric or abiotic phases of carbon dioxide are closely interlinked. The multiplicity of the pathways along which carbon can flow is typical of the biosphere cycles in general and it provides a well-buffered system with adequate feedback mechanism to ensure an adequate supply of carbon. It is significant that all phases of the CO_2 cycle yield carbon dioxide at some time and carbon dioxide is the raw material for them. The result is that, inspite of its relatively low concentration in the atmosphere (about 0.03%), carbon is virtually available in a form in which it can be used by living organisms.

7.4.3 Nitrogen Cycle

7.4.3.1 Basic Ideas

Nitrogen occupies a special place from the point of view of human life because it is a sensitive indicator of the quality of health in human beings. A lack of nitrogen in the form of protein deficiency quickly leads to a weakened condition and poor health. Nitrogen is a key element in many essential molecules like proteins, nucleic acids, enzymes, amino acids vitamins and hormones. Nitrogen is also a major constituent of the atmosphere, comprising approximately 79% of it; but the paradox is that, in its gaseous phase, nitrogen is not available to most life. Before it can be utilized by organisms, nitrogen must be converted to a chemically usable form. To be used biologically, the free molecular nitrogen in the atmosphere has to be fixed and fixation requires an input of energy. In the first step, molecular nitrogen has to be split into two nitrogen atoms ($N_2 \rightarrow 2N$). The free nitrogen atoms then must be combined with hydrogen to form ammonia, with the release of some energy, $2N + 3H_2 \rightarrow 2NH_3$.

7.4.3.2 The N_2 Cycle

The major points in the nitrogen cycles may be described as follows :

(1) Plants and animals continuously produce proteins, which are organic compounds containing nitrogen.

(2) Nitrogen is abundant in the form of N_2 gas. Unfortunately, the nitrogen in gaseous form cannot be used directly by most organisms.

(3) A very small number of plants like peas, beans and grams can convert or fix the nitrogen in gaseous form directly into organic compounds primarily in the form of proteins. These plants achieve the fixation of nitrogen with the help of certain bacteria living in the root nodules of these plants.

(4) Most of the nitrogen in living organisms do not enter into their bodies directly from the atmosphere. Instead, another group of bacteria in the soil and water can convert gaseous nitrogen to inorganic nitrogen in the form of nitrate. Plants then obtain this nitrate from the soil. Since nitrogen is essential for photosynthesis, the amount of nitrate in the soil can regulate plant growth.

(5) In the total cycle, about 4-7 tonnes of nitrogen per hectare is added to the soil each year. There is some loss from the soil through the leaching of nitrates into fresh water courses (streams, rivers and lakes) and the seas. However, the nitrogen cycle is balanced by all these processes. As a result, the concentration of nitrogen in the atmosphere remains more or less constant.

(6) When herbivores eat plants, some of the organic nitrogen is transferred to these animals and eventually, to other animals (the carnivores).

(7) When plants and animals die, their organic nitrogen is converted by decomposers to inorganic nitrogen in the form of ammonia (NH_3), which is converted by another group of bacteria to nitrite (NO_2).

(8) Ammonia can also be converted to atmospheric nitrogen (molecular nitrogen, N_2), or back to nitrate (NO_3), which can then be taken up by plants to begin the cycle again.

(9) Some nitrogen is lost from the cycle when the soluble nitrates are removed from the soil by water, carried away by rivers and deposited as deep-sea sediments.

(10) Under natural conditions, nitrogen lost from ecosystems by denitrification, leaching, erosion, etc is balanced by biological fixation of atmospheric nitrogen. Terrestrial and aquatic ecosystems constitute a dynamic equilibrium system both chemically and biologically in which a change of nitrogen in one phase affects the other and the balance is maintained.

7.4.4 Oxygen Cycle

7.4.4.1 Basic Ideas

After nitrogen, oxygen is the second most abundant element in the atmosphere. It constitutes approximately 20.95% V/V (volume per volume) of the atmosphere. Oxygen is an essential element for all living organisms. We inhale oxygen (O_2) during respiration and release carbon dioxide (CO_2). This carbon dioxide is utilised by plants during photosynthesis and the plants release oxygen. Thus the animals and plants maintain the balance of oxygen in the atmosphere. Ozone (O_3) is another form of oxygen, which forms a layer in the stratosphere and protects life on the earth from the solar ultraviolet radiation. Oxygen dissolved in water is the main source of oxygen for aquatic organisms.

7.4.4.2 The O_2 Cycle

Oxygen is needed by most plants and animals including human beings for aerobic respiration or enzymatic oxidation of organic food to sustain growth and general metabolism. Oxygen is absorbed both by plants and animals from the environment during their aerobic respiration, but released by green plants during photosynthesis, thereby setting up the oxygen cycle. There is also a continuous exchange of oxygen between the atmosphere and all water bodies on the earth. As a result, the total amount of oxygen in the biosphere remains fairly constant, so that the oxygen cycle is stable.

The oxygen cycle is based on the exchange of O_2 among various segments of the environment, viz., atmosphere, hydrosphere, lithosphere and biosphere. In photosynthesis, oxygen is freed from the water molecule. This freed oxygen is reconstituted into water during the aerobic respiration of plants and animals. Some part of the atmospheric oxygen that reaches the higher level of the stratosphere is converted to ozone (O_3) by high-energy ultraviolet radiation from the sun.

7.4.5 Water Cycle

7.4.5.1 Basic Ideas

The constant circulation of water from the oceans to the atmosphere as water vapour and then back again to the oceans with temporary residence in living organisms, fresh water bodies, ice and snow accumulations, or as ground water is known as water (or hydrologic) cycle. The water cycle is an intricate combination of evaporation, transpiration, air mass movement, condensation, precipitation, run off, percolation and ground water movements. The greater part of atmospheric moisture, which eventually falls to earth as precipitation (rains), comes from the oceans. Some water takes a short cut in the water cycle and enters air (atmosphere) directly through evaporation and transpiration (from soil and vegetation) without returning to the oceans first. The processes of evaporation, condensation and precipitation are essentially climatic and their functions in the water cycle would be quite simple if it were not for the constant motion in the atmosphere.

This cycle requires heat as well as moisture. Where heat and moisture are abundant (e.g., in rainy, tropical countries), the water cycle is very active. In dry climates, however, an essential part of the cycle, viz., moisture, is lacking. In very cold climates, on the other hand, the energy (i.e., heat) to operate the water cycle is limited. The practical value of an understanding of the water cycle comes from the fact that it is this cycle that makes water available to us on land for its various uses.

7.4.5.2 The Water Cycle

Under the influence of solar radiation, water in oceans, seas, rivers, lakes and other water bodies is constantly vaporized. The process of vaporization, which sends water into the atmosphere in the form of water vapour, can occur in many different ways. In the first place water evaporates directly from the surface of seas, lakes, rivers, glaciers, snow fields, or straight out of the ground. A second process to convert water into vapour is respiration. All living organisms emit carbon dioxide and water vapour into the atmosphere in the process of respiration. A third process is the emission of water vapour by the combustion of organic materials. However, the largest proportion of water vapour in the atmosphere comes directly from the oceans.

The major pathway of the water cycle is an interchange of water between the earth's surface and the atmosphere through precipitation and evaporation. It is a steady-state cycle since the total precipitation is balanced by the total evaporation. Although there is a greater precipitation on land than on oceans, the run-off from the land to the seas and oceans compensates for this imbalance.

The water cycle is a continuous natural process, which helps in the exchange of water among the atmosphere, the land, the seas and the living plants and animals. Approximately one-third of the solar energy received on the earth is utilized to drive the water cycle. Water is precipitated on all land and water surfaces in the form of rain or snow. The water precipitated on land surfaces seeps into the soil as ground water. The ground water does not remain stationary but moves in various directions. The ground water moves up above the water table by capillary action and thereby maintains a continuous supply of water to the surface layer of the soil, where it is absorbed by plant roots in the absence of rain.

The water cycle over the oceans is extremely simple. On the land, however, the water cycle is more complex. The reason is that several routes are open to the water that precipitates on land, viz., (1) direct evaporation, (2) transpiration, (3) entry of the water into the ground water system and (4) run-off. These routes of the water cycle on land can be divided into three major categories :

(1) Evapotranspiration (the rapidly cycling water).
(2) Surface run-off (the less rapidly cycling water).
(3) Ground water (the very slowly cycling water).

7.4.6   Some Other Cycles

7.4.6.1   Sedimentary Cycles

Mineral elements required by living organisms are obtained initially from inorganic sources. Available forms of these elements occur as salts dissolved in soil, water, streams, lakes and oceans. The **mineral cycle,** in general, varies from one element to another but essentially it consists of two phases, viz., (1) the **salt solution phase** and (2) the **rock phase**. Mineral salts come directly from the earth's crust by weathering. Among these salts, the soluble ones enter into the water cycle and with water, they move through the soil to streams, rivers and lakes and eventually reach the seas and oceans, where they remain indefinitely. Other salts are returned to the earth's crust through sedimentation. They become incorporated into salt beds, silts and limestones and enter the cycle again after weathering. The minearal cycles are also known as **sedimentary cycles.**

Plants and animals fulfil their mineral requirements from the mineral solutions in their environment. Many animals acquire the bulk of their minerals from the plants and animals they consume. After the death of living organisms, the minerals are returned to the soil and water through the action of the decomposers.

There are many different types of mineral or sedimentary cycles depending on the type of elements involved. Two important sedimentary cycles, viz., the sulphur cycle and the phosphorus cycle are significant for ecosystem.

7.4.6.2 The Sulphur Cycle

Sulphur, like nitrogen, is an essential part of proteins and amino acids. It exists in a number of states such as elemental sulphur, sulphides, sulphites, sulphates, etc. The sulphur cycle has both sedimentary and gaseous phases. The sedimentary phase of the sulphur cycle is a long-term phase. In this phase, sulphur is tied up in organic and inorganic deposits. It is released from these mineral deposits by weathering and decomposition and is carried to terrestrial and aquatic ecosystems in a salt solution. Sulphur enters the atmosphere from several sources such as combustion of fossil fuel, volcanic eruptions, gases released by decomposition of dead plants and animals, etc. Initially, sulphur enters the atmosphere as hydrogen sulphide (H_2S), which quickly oxidizes into sulphur dioxide (SO_2). Atmospheric sulphur dioxide, which is soluble in water, is carried back to earth in rainwater as weak sulphuric acid (H_2SO_4). Sulphur in a solution form, mostly as sulphate (SO_4), is absorbed by plants through their roots, where it is incorporated into certain organic molecules like proteins and some amino acids. From the producers (i.e., green plants), sulphur in amino acids is transferred to the consumers.

7.4.6.3 The Phosphorus Cycle

In contrast to sulphur cycle, the phosphorus cycle has no atmospheric (or gaseous) phase. It occurs naturally as phosphate (PO_4), or one of its analogues (e.g., HPO_4 or H_2PO_4), as soluble inorganic phosphate ions, as part of soluble organic molecules, as particulate phosphate or as mineral phosphate. The ultimate sources of phosphate in the ecosystems are the crystalline rocks. As they are eroded and weathered, phosphate is made available to living organisms, generally as ionic phosphate, which is introduced into the autotrophic plants through their roots. From autotrophs, phosphorus is passed along the grazing food chain to other animals in the same way as nitrogen or sulphur, with excess phosphorus being excreted by animals in the faeces. Phosphates can also be released as particulate matter from forest and grassland fires.

After the death of plants and animals, phosphorus is liberated from large organic molecules, by the action of decomposers, in the form of inorganic ionic phosphate. In this form, it can be immediately taken up by the autotrophs to begin the cycle again, or it can be incorporated into a sediment particle. The sedimentary phase of phosphorus cycle remains comparatively slow compared to its organic phase.

7.5 MAN AND ECOSYSTEMS

7.5.1 Man's Ecologic Relations

Man is an integral part of nature and dependent upon it for food, clothes, shelter, several raw materials and energy. In contrast to other animals man has succeeded in surrounding himself with a complex culture of tools, artifacts, customs and institutions, which stand between him and his natural environment. Man has achieved an extraordinary freedom and control over his environment. In order to describe the environment, the ecologist begins with the earth as a physical system with a continuous flow of energy. As a consequence of this energy flow, various chemical materials of the earth's surface circulate in biosphere cycles. Most of the energy input of this system comes from the sun.

The biosphere comprises numerous communities (like forests and grasslands), which interact with their nonliving environments to form a series of discrete ecosystems within the framework of the world system. Each community is held together by a complex web of food chains and other relationships in a state of such mutual dependence that any change in one component is likely to cause changes throughout the rest of the system. Ecologist regards this balance of nature not as a static condition but as a **dynamic equilibrium**, gradually achieved in the course of an immensely long organic evolution. In such a system, each type of organism has its characteristic ecologic **niche**, defined not only in terms of its particular habitat requirements but also in terms of its functions in the community, such as those of predator or prey, host or parasite, or producer, consumer or decomposer of organic materials. Each plays some role in the maintenance of the ecosystem and its influence is generally held in check by its interaction with other components of the ecosystem, both living and nonliving. This provides the ecosystem with numerous built-in **regulatory mechanisms** and permits a very considerable degree of ecologic stability or **homeostasis.**

The perception that man lives in close relationship with all living organisms, both animals and plants, is quite old. Recent developments in science have confirmed the **inter-dependence** of all lives, between themselves and with the physical environment of land, water, air and solar energy. A new dimension began to appear in ecology in the last few decades of the twentieth century as the impact of man and his activities on nature came to be noted. The word "**environment**" was given an additional meaning during this period to express the broader concept of interaction of human activities on the ecosystem and the impact of the changes in ecosystem on man himself.

7.5.2 Dominance of Man

Man has become a major force in every ecosystem with which he has come in contact. The age of the earth is estimated to be about 4.5 billion years. Life on earth has probable existed for one billion years. According to oldest fossils available, man has existed on earth for about one million years.

A sudden shift in the pattern of human living occurred about 2,000 years ago, when man adopted a sedentary mode of life based on the domestication of animals and the cultivation of plants, thus changing from a food gatherer to a food producer. As his technology increased, man gradually expanded his use of non-biological sources of energy, with the development of sails and later water mills (around 100 B.C.) and wind mills (around 900 A.D.). Although coal has been an important source of energy from the twelfth century, large-scale consumption of energy has occurred only within the last two hundred years, with the harnessing of steam, electricity, petroleum, natural gas and, most recently, of atomic energy. If man's rate of energy consumption is used as an index, the growth of his ecologic dominance can be seen to have followed an **exponential growth.**

7.5.3 Man's Alteration of Nature

In course of developing his present ecologic niche, man has made significant alterations in the organization of the natural world. Among the most important of these alterations are the changes in the vegetational pattern. In Europe and North America and to a lesser extent in Asia, Africa and South America, there has been extensive removal of forests and woodlands, for fuel and timber products, as well as drainage of swamps and other wetlands. These have been replaces by orchards, pastures and fields widely sown in monocultures or single-species crops. Wide-spread abandonment of land after short periods of intensive cultivation has resulted in large areas of shrub vegetation. Moreover, the importance of fire as an environmental factor has increased greatly since man learned to make use of it. Evidence suggests that it has played an important role in the development of many tropical and temperate grasslands.

Agricultural and forest removal practices have altered patterns of soil development and rate of evapotranspiration. They have also increased the rate of soil erosion over millions of acres, thus contributing to the rapid deposition of silt in many rivers and in both natural and artificial reservoirs. The resulting reduction in total quantity of plant life has promoted instability in local microclimates. It may also be responsible for changes in the composition of the atmosphere, e.g., the 10% increase in carbon dioxide that has been observed since 1900 at mid-latitudes in the northern hemisphere. Construction of large dams, artificial lakes and irrigation canals has considerable effect on the pattern of water distribution and availability. The growing network of highways has constricted the natural landscape into increasingly smaller pieces.

Equally important are the more direct effects which man has on the world's fauna and flora. He has caused the extinction of many species. Accidentally or deliberately, man has introduced many thousands of plant and animal varieties into all the continents and most of the remote oceanic islands. The development of rural and urban habitats has brought man into unwanted contact with rats, house-flies, cockroaches and many other species, which have invaded his shelter. Thus, man has caused the most drastic biological disturbances the earth has experienced so far.

7.5.4 Signs of Eco-Imbalance

There is considerable evidence to indicate that diversity of the habitat and richness of **biota** (species of all plants and animals occurring within a certain area or region) are positively correlated with the stability in natural communities. Man's net influence has been to reduce the biodiversity and to impoverish local flora and fauna, producing simpler and more uniform conditions that are conducive to fluctuations and imbalance. As a result, man has created many serious problems which has affected the continuation of his ecological status. The uncontrolled discharge of smoke and fumes from industrial stacks and domestic chimneys and the exhausts of vehicles have almost turned some of our large cities into gas chambers. Pollution has also resulted from the development of chlorinated hydrocarbons like DDT and chlordane, organophosphates like malathion and parathion, and other synthetic pesticides to kill pests, insects and weeds.

7.5.5 Limits to Growth

Growth of population and economy has created an increasing demand upon the environment for raw materials and natural resources. Although some regions of the world show food surpluses at present, other regions have serious deficits in food production. On the whole, approximately two-third of the world's present population have an inadequate diet. To secure adequate food supply to all mankind in the near future, a global increase of food production of about 2.25% per year will be required; but sustained rates of increase of this magnitude have been achieved only in North America. Water has always been a limiting factor for man because of its uneven distribution and the difficulty of preventing its loss by evaporation and run-off. Man's need and use of water have increased steadily, especially in the highly industrialized countries. Many have predicted increased hunger, disease and misery, leading even to the outbreak of nuclear war, with recovery (if any) to an agrarian level. Certainly, there are reasons behind the problems of maintaining a satisfactory standard of living without exhausting the supply of natural resources and of achieving a stable ecological position in the world ecosystem.

7.6 CAUSES FOR ECO-IMBALANCE

7.6.1 Role of Human Activities

Many of the large-scale human activities, agricultural, industrial or transport tend to alter the natural balance of living organisms in an ecosystem and create an imbalance in it. Such a disturbance in the ecosystem can, in turn, affect human societies in a number of ways both immediately and in the long term. In a sense, we have rediscovered the ancient wisdom of our land (*the intimate relationship between Prakriti and Purusha*) that if the ecosystem is destroyed or seriously damaged, this has the potentiality of jeopardizing human welfare. The effects of the activities of man on the global ecosystem are compounded by the factors of his increasing technological ability, large increases in population and greater consumption of material and energy.

It is obvious that the main reason for imbalance in ecosystem of the world is man's dominance in the global ecosystem and his thoughtless alternation of it to satisfy his need and greed. Some specific causes for eco-imbalance are as follows :

(1) Changing hydrology.
(2) Large-scale use of fertilizers.
(3) Large-scale use of pesticides.
(4) Environmental pollution.
(5) Changes in physical and biological properties of earth's surface.
(6) Acid rain.
(7) Green-house effect.
(8) Forest fires.
(9) Overgrazing.
(10) Clearance of forests.
(11) Plant and animal breeding and their introduction into new areas.
(12) Loss of genetic resources.
(13) Mining activities.
(14) Industrialisation.
(15) Urbanisation.

These are briefly discussed in the following sections:

7.6.2 Changing Hydrology

Irrigation is a major consumer of water in many countries. When river water is used for irrigation by constructing canals, the rate of water flow in the river is reduced and the quality of water down-stream may decline. Building large dams for irrigation and power generation changes the ecosystem of the area impounded for the reservoir. In the impoundment area, a large number of trees may be felled, fertile land may be submerged under water and a large human population may be displaced. In addition, irrigation increases the water vapour contents of the surrounding air, alters the surface run-off characteristics of the land and sometimes, increases the concentration of pesticides in the underlying ground water.

7.6.3 Large-Scale Use of Fertilisers

Plants rarely use more than 50-60 percent of the nitrogen chemical fertilizers. The residual nitrogen in the form of nitrate is likely to pollute ground and surface waters, causing over-enrichment of nutrients and algal blooms in rivers and lakes. Some of the extra nitrates may also be converted into nitrogen oxides by the action of certain soil bacteria. In order to reduce the adverse ecological impact of the large-scale use of fertilisers, researches are being conducted currently to determine the extent of fertiliser use and minimize environmental pollution, while maximizing food production. A novel method is to employ fertilizers that release the nutrients slowly. Such fertilisers have already been developed, but they are very expensive. Another approach is to improve the efficiency of biological nitrogen fixation and to extend it to more crop species.

7.6.4 Large-Scale Use of Pesticides

Pesticides are the chemicals used to eradicate pests. Pesticides are classified as insecticides, rodenticides, fungicides, nematicides, herbicides and so on. However, none of these chemicals are so specific as to kill only the target species. As a result, they pose hazards also to other organisms including man. The problems associated with the well-known pesticide DDT became evident in the early 1960's with large-scale accumulation of DDT residues found in soil, crops, animal tissues and even in humans. This lead to the banning of DDT in most of the industrialized countries.

There are other types of pesticides such as *organophosphates* and *carbamates*, which are less persistent than organochlorides; but they are more water soluble and can leach into ground water. They are also more toxic to birds, animals and human beings.

7.6.5 Environmental Pollution

The major causes of environmental pollution are industrialization, urbanization and motorization. In many countries a large number of industries have been set up inside or near large cities. The stack emissions from these industries pollute the air that man breathes. The industrial effluents pollute the streams, rivers and oceans.

A consequence of industrialization, especially in developing countries, is the large-scale migration of rural population to the industrial centres in large cities, leading to their overpopulation. This results in congestion within the city, creating slum conditions, infections and water and sewage problems. The neo-rich people leave the city and move to suburbs. In the process, they fell trees, drive away wild animals from their natural habitat or kill them, drain pond and other natural water reservoirs and clear forests for constructing residential complexes. The consequence of these is interference with the ecosystem and degradation of environmental quality.

7.6.6 Changes in Earth's Surface

The atmosphere and the ecosystem can be affected both by human activities and natural changes in the physical and biological properties of the earth's surface. Some examples of such changes are as follows :

(1) Alteration in forest cover, irrigation or desertification; snow clearance on land; ice clearance in shipping lanes; and construction of towns, roads or airports. All of these can change the fractions of solar radiation reflected and absorbed by the ground surface.

(2) Modifications (including deforestations and swamp drainage) that reduce evaporation from an area and alter the amount of energy available for evaporation, thereby changing the energy balance at the earth's surface.

(3) Alterations in the hydrologic or water cycle through the construction of dams for irrigation and power generation, overgrazing, diversion of rivers, etc.

(4) Changes in the strength of natural sources and sinks of trace substances. Forests, for example play an important role in the carbon (or carbon dioxide) cycle. A change in the forest cover, therefore, affects the concentration of atmospheric carbon dioxide.

7.6.7 Acid Rain

In common language, *"acid rain"* means the presence of good amount of acids in rain water. Acid rain is one of the major effects of air pollution. Large amounts of sulphur dioxide and oxides of nitrogen are released into the atmosphere through the burning of fossil fuels (oil and coal) and fuel wood. These oxides (SO_x and NO_x) react with the atmospheric water vapour to produce sulphuric acid (H_2SO_4) and nitric acid (HNO_3), which then return to the earth's surface with rain water. Acid rain poses a major threat to ecological balance with potential for both macro and microbiological effects on plants and animals. Acid rain not only affects crops, forests and aquatic ecosystems adversely, but it also leaches exposed rocks, there by damaging ancient monuments such as the Taj Mahal. Since many famous structures have been constructed from soft rocks, acid rain has already damaged many ancient monuments of great historical value.

7.6.8 Green-House Effect

The atmosphere of the earth is said to act like a "green-house" and carbon dioxide is one of the major components of the atmosphere which stimulates the *"green-house effect"*. The green-house becomes warm because glass is transparent to visible light but practically opaque to infrared radiation. As a result, the sunlight can pass through the glass walls of the green-house and can get absorbed by the soil inside it. It is then reemitted as heat rays (infrared radiation), which cannot pass through the glass walls. As a result, the temperature inside the green-house rises above the ambient temperature.

In a similar way, the atmosphere of the earth bottles up solar energy due to the presence of carbon dioxide, water vapour, methane, ammonia and a few other gases (the so-called "green-house gases") in the atmosphere. These gases transmit the visible light from the sun and allow it to reach the earth's surface, but reflect (by absorption and subsequent reemission) the heat rays (infrared light) from the earth's surface. As a result, a small change in the composition of the atmosphere can have a marked effect on the terrestrial temperature.

The rising proportion of carbon dioxide is causing great concern because the green-house effect will result in higher average temperature on the surface of the earth (the so-called *"global warming"*). If this happens, there will be wide-spread climatic change with possibly disastrous consequences. The rising temperature could melt the polar ice caps, submerging much of low-lying land mass and many coastal cities (like London, New York, Mumbai, Kolkata and Chennai) under sea water. Similarly, fertile land may be turned into desert and agricultural production may fall drastically due to global warming.

7.6.9 Forest Fires

Fire is still employed to help man hunt wild animals, clear forests and obtain charcoal for fuel. Repeated forest fires, combined with overgrazing, can seriously degrade the environment and harm the ecosystem.

7.6.10 Overgrazing

Environmental degradation and ecological damage arising from overgrazing are wide-spread in arid and semi-arid regions of the Near-Eastern countries, Central Asia and the Mediterranean basin for many years. One important consequence of overgrazing is desertification (i.e., a process of reduction or elimination of the productive capacity of land that leads ultimately to desert-like formation). About 95% of the land in the arid and semi-arid regions of the world are in the processes of desertification.

7.6.11 Clearance of Forests

Forest represents at least 6 billion hectares for natural vegetation and several billion hectares more could easily support woody vegetation. It has been estimated that, by the mid-twentieth century, mankind reduced the world's original forest area by at least 33%. Man has converted forests to grasslands and croplands. In addition to the clearance of forests for agriculture and animal husbandry, the forests are being destroyed for timber and fuel-wood.

7.6.12 Plant and Animal Breeding

Sometimes plants have been introduced in new habitats which have brought nutritional value to the diets of millions of people. Similarly, animals beneficial to mankind have been added. Unfortunately, many introductions of new plant and animal species have been without adequate forethought and without the collection of ecological data on the probable impact on native plants and animals. For example, the house sparrow, which was introduced in North America, soon became a serious crop pest. In Australia, the introduction of European carp now seriously threatens the environment of native fish species in island waters, mainly because it increases the turbidity of water, which reduces the productivity of aquatic plants. The best known case in Australia is the introduction of rabbits from Europe. The rabbits have multiplied rapidly since they have no natural enemies in Australia, and have caused extensive damage to the native flora and fauna.

7.6.13 Loss of Genetic Resources

Since the 1970's, great concern has been expressed over the survival of thousands of plant and animal species. Hunting of animals for food or sport has taken its toll on a number of species, and so too the excessive plant collection for medicinal or other purposes. According to a study by the International Union for Conservation of Nature and Natural Resources, about 1,000 species of birds and mammals are currently threatened with extinction. Destruction of habitats is one of the major reasons for the loss of species. Such a loss causes eco-imbalance.

7.6.14 Mining Activities

Mineral extraction and processing has a wide range of ecological impacts, which can be divided into four main categories :
   (1) Impact on land.
   (2) Impact on atmosphere.
   (3) Impact on water.
   (4) Impact on socio-economic environment of local people.

   Direct effects of mining on the landscape such as surface disturbances and generation of wastes tend to be roughly proportional to the quantity of minerals extracted. Reclamation of land disturbed by mining activities is an important factor in reducing the environmental damage caused by mining wastes. If surface mining is extended more and more to areas with fragile ecosystems, the rehabilitation of land after mining becomes a severe problem. Progress in rehabilitation is expensive and slow in areas where soil cover is thin, the over-burden is high in acidity or salinity and rainfall is sparse.

   Air pollution generated from mining and ore processing activities creates serious environmental and health problems.

7.6.15 Industrialisation

Industrialisation is necessary to provide the basic necessities and comforts of life to our growing population. However, improperly planned industrial-ization has created serious problems of environmental pollution and ecological imbalance all over the world.

7.6.16 Urbanisation

Unplanned urbanisation gives rise to many socio-economic, environmental and health problems. For example, slum settlements arise as part and parcels of a city and are essentially the result of acute shortage of housing in cities and towns. People not able to find a dwelling tend to occupy any vacant land outside and inside the city, where they build a dwelling with any available material. This is done in a haphazard manner with no civic facilities of water supply, drainage, roads, transport and other amenities. A slum settlement becomes a burden not merely due to the poor quality of structures, but also on account of the environment of insanitation that leads to several social evils. Slum expansion leads to destruction of natural ecosystems surrounding the towns and cities.

7.7 EFFECTS OF ECO-IMBALANCE

7.7.1 Impact of Eco-Imbalance

Some important impacts of polluted environment and imbalanced ecosystems are as follows :

(1) Ill effects on human health.
(2) Soil degradation.
(3) Desertification.
(4) Genetic resource depletion.
(5) Contamination of food.
(6) Trace substances in the troposphere.
(7) Depletion of the stratospheric ozone layer.
(8) Global warming.

These are briefly discussed below :

7.7.2 Ill Effects on Human Health

Polluted air, water and land generate many harmful chemical and biological agents that have a negative impact on human health. A wide range of communicable diseases can be spread through the elements of environment polluted by human and animal waste products. This is clearly evidenced by the plagues of the Middle Ages, when the disease spread through rats that were fed on contaminated human waste. Although major diseases transmitted via the environment have been almost eliminated in developed countries through immunization and sanitation programmes, no country is totally immune from outbreaks of environmentally transmitted diseases, as the outbreak of SARS (Severe Acute Respiratory Syndrome) in 2003 in many countries clearly proved.

7.7.3 Soil Degradation

The protection of soil against the hazards of degradation is essential if the productivity of soil has to be sustained. Soil degradation has many causes, but the immediate concerns are improper land use, soil erosion, acidification, salinisation, water-logging and chemical degradation. Soil erosion is the washing or blowing away of the surface soil. Erosion may take place under natural conditions, but it is greatly increased when human activities cause disappearance of the protective cover of natural vegetation. Acidification and salinisation directly reduce soil fertility. They may be caused by acid rain and accumulation of water soluble salts in the soil. Chemical degradation of soil may occur if the nutrients in the soil are leached out or not replenished to maintain soil fertility and if the soil is contaminated by harmful chemicals like DDT and radioactive substances. Soil erosion is a global problem,

7.7.4 Desertification

The term **"desert"** encompasses a wide range of environmental complexes:
 (1) Rainless deserts, where rainfall is not an annually recurring event.
 (2) Run-off deserts, where the annual rainfall is low (less than 100 mm) and variable.
 (3) Rainfall deserts, where the rainfall is insufficient for crop production (100-200 mm).
 (4) Man-made deserts, parts of the semi-arid areas (rainfall 200-350 mm) that have been transformed into deserts due to man's over-exploitation of the land.

Desertification results from the combined effect of two factors; severe recurrent droughts and human over-exploitation of drylands. The cures for desertification have been known for a long time. They consist of the reverse processes, i.e., biological recovery of environmental conditions, naturally or artificially induced. Considerable experience in combating desertification has been acquired by the US, Australia and Israel but corrective measures are expensive though net benefits would certainly exceed the costs.

7.7.5 Genetic Resource Depletion

The genetic material contained in the domesticated varieties of crop plants, trees, livestock, aquatic animals and microorganisms is essential for the breeding programmes in which continued improvements in yields, nutritional quality, flavour, durability, pest and disease resistance, responsiveness to different soils and many other qualities are achieved. Because of intensive selection for high performance and uniformity, the genetic base of much food production in modern times has grown dangerously narrow. For example, only four varieties of wheat produce 75% of the crop grown in the Canadian Prairies and more than half of the Prairie wheatlands are devoted to just a single variety. Similarly, 72% of the potato production in the US depend on only four varieties. Unfortunately, while the genetic base of the world's crops and other living resources is narrowing rapidly, the means by which this dangerous situation can be corrected are being destroyed due to ecological imbalances created by human activities. Many wild and domesticated varieties of crop plants (such as wheat, rice, millet, beans, tomatoes, potatoes, bananas, limes, oranges, etc) are already extinct and more are in danger. Useful breed of livestock is also at risk. Of the 145 indigenous cattle breeds in Europe and the Mediterranean region, 115 are threatened with extinction.

7.7.6 Contamination of Food

Chemical contaminants reach food and livestock feed from many sources. Pesticides used in farming often find their way into crops. In addition, veterinary drugs and animal growth promoting chemicals may pass into meat and dairy products like milk and butter. Some food preservatives like sodium nitrite, chemicals and materials contained in food packaging may also enter the packaged food. Crops may be chemically contaminated by the airborne deposition of industrial emissions or by industrial effluents. The current trend to centralize food processing, handling and distribution of food and the greater reliance on large storage facilities may aggravate some of the above problems.

Considerable attention has been paid to monitoring the amount of pesticides in food, establishing standards for maximum permissible concentrations and finding means of keeping the levels of pesticide residues low. Government regulations and monitoring, coupled with strict enforcement of these regulations, have led to a decline in the concentration of pesticides in many countries. In India, however, pesticides in concentrations higher than recommended standards have been found in many food items, soft drinks and even mineral water. In addition to pesticides, highly toxic heavy metals like mercury, cadmium, lead, arsenic and selenium are among contaminants of food. Many of these enter the soil with urban wastes. From the soil, the heavy metals enter the food crops and fodder. Sewage sludge, raw sewage, manure and farm wastes used as soil conditioners often lead to elevated concentrations of metals in the food supply. Other heavy metals enter the food supply from discharges of industrial effluents into water course or by direct disposal on the soil. Metals like mercury and tin are present in pesticide formulations. In addition, metallic dusts especially those containing lead are locally deposited on the leaves of food crops, vegetables and fruit plants. The plants accumulating metals may be used as food or feed source and hence enter the human food chain.

Fungicides containing mercury have been used widely in agriculture as a seed treatment to control pathogens affecting seed germination or subsequent growth of seedlings. Many disastrous causes of human poisoning from consumption of seed grains treated with fungicides have been reported, the worst incident was the death of more than 500 people in Iraq in late 1971 and early 1972. Birds have also been killed by eating treated seeds. As a result, fungicides containing mercury have been banned in many countried and have been curtailed in some countries.

7.7.7 Trace Substances in the Troposphere

Man introduces trace substances into the troposphere (the lower portion of the atmosphere, usually extending up to about 11 km from the earth's surface) by :

(1) Emitting gases and particulates (especially from the combustion of fossil fuels, domestic and industrial sources and transport).
(2) Forest and grassland fires.
(3) Ploughing and overgrazing, thus releasing dust that rises in dry and windy weather and augments the dust already present in the atmosphere.
(4) Releasing particulates that act as nuclei for condensation and freezing.

The most important trace substances introduced into the troposphere are the traditional pollutants of towns and cities, viz., suspended particulates, oxides of sulphur and nitrogen, carbon monoxide, hydrocarbons, hydrogen sulphide, lead, mercaptans and fluorides. Some of these pollutants are long lived toxic substances which are transported by the atmosphere to areas where they may accumulate in organisms, soils, water bodies or glaciers. Others are photochemically reactive gases (like nitrogen oxides and hydrocarbons) whose reactions in the presence of sunlight produce oxidants like ozone. Oxides of sulphur and nitrogen also produce sulphate and nitrate aerosols, which cause acid rain. Other gases (mainly carbon dioxide, but also chlorofluorocarbons, nitrous oxide, methane and ammonia) absorb the outgoing radiation from the surface of the earth, thereby leading to global warming. Aerosols and suspended particulates may also affect the radiation balance of the atmosphere and influence the climate. Finally, some gases such as chlorofluorocarbons reach the stratosphere and cause the depletion of ozone layer.

7.7.8 Depletion of the Stratospheric Ozone Layer

Following the predictions in 1974 that chlorofluorocarbons as well as some other gases (methane, ammonia and nitrous oxide), diffused to the stratosphere, would lead to the depletion of the protective ozone layer, a group of experts prepared a World Plan of Action on the Ozone Layer. Under this Plan, a Co-ordinating Committee on the Ozone Layer was established, which subsequently produced assessments of the ozone layer and its impacts.

7.7.9 Global Warming

Although carbon dioxide is almost completely transparent to visible light, it strongly absorbs and reradiates the infrared radiation (heat rays). Thus carbon dioxide acts like the glass of a green-house and, on a global scale, tends to warm the air in the lower levels of the atmosphere. This phenomenon is known as the "**green-house effect**". In addition to carbon dioxide, water vapour and ozone also absorb the infrared radiation emitted by the surface of the earth and help keep the earth warm. There is enough evidence to show that the temperature of the entire earth has risen during the recent decades. For example, glaciers in both hemispheres are receding. There occurred an increase in the mean annual air temperature of about 0.5°C during 1885-1940. The highest increase in the annual mean temperature occurred in the Northern Hemisphere between 40°N and 70°N latitudes, where the average rise in temperature was 0.9°C and astonishing increase in the average winter temperature of 1.6°C.

Although global warming subsided after 1940, warming of Northern Europe and North America continued between 1940 and 1960. More than 40% of the total increase in carbon dioxide content of the atmosphere from combustion had taken place during this period; but the mean annual air temperature worldwide as well as for the Northern Hemisphere decreased by 0.1°C. It is evident, therefore, that other causes of climatic change have been atleast as important as the carbon dioxide.

7.8 REMEDIES FOR ECO-IMBALANCE

7.8.1 Restoring the Eco-balance

Some important strategies for removing the ecological imbalance in the world ecosystem are as follows :

(1) Use of biotechnology.
(2) Development of alternate energy sources.
(3) Conservation.
(4) Control of thermal pollution.
(5) Recycling and substitution.
(6) Development of eco-friendly technologies.
(7) Development of low-cost technologies.
(8) Use of alternate resources.

7.8.2 Use of Biotechnology

Biotechnology offers the prospects for the production of valuable substances, or the removal of undesirable materials, with the help of microorganisms and their enzymes. Man has always used microbes in fermentation industries. In the early twentieth century, these industries developed further to produce chemicals like ethanol and acetone from starch and molasses. The production of antibiotics, enzymes and vaccines followed later. Biotechnological methods have also been used on a large scale in aerobic and anaerobic treatment of wastes.

Recent advances in genetic engineering permit the transfer of genetic material from a variety of species with desired characteristics into selected recipient microbes. These microbes then indefinitely produce new individuals with desired characteristics. Examples of the new approach are the large-scale ethanol industry in Brazil, the wide-spread digesters for producing biogas, the production of selected nitrogen-fixing bacteria, the production of pest pathogens (for use in biological pest control) and the expansion of fodder yeast production from hydrocarbons. The Imperial Chemical Industries in the United Kingdom have already set up a factory producing fodder supplement from methanol.

The potential of biotechnology as expanded by genetic engineering is judged to be very high. Already, a bacterial strain has been patented that is capable of degrading a wider range of oil fractions than any other strain now known. Biotechnological productions of insulin and certain vaccines have been shown to be possible. Much work has been devoted to the implementation of nitrogen-fixing genes into cereals, which might reduce reliance on nitrogenous fertilizers. Further research is aimed at biotechnological production of the hydrogen gas (for fuel) and artificial photosynthesis with the help of enzymes.

7.8.3 Development of Alternate Energy Sources

A major preoccupation of various governments and scientists at present is the question of energy alternatives, particularly the environmental aspects of the expanding nuclear industry and increasing utilization of coal in many countries. This has lead to the growing concern about the problems of acid rain and global warming. An important common element of both the acid rain and the carbon dioxide problem is the biogeochemical cycling of trace substances through the environment. For example, the atmospheric components of both the sulphur and nitrogen cycles are limited, which means in practice that the control of sulphur emissions alone would not necessarily have the expected effect.

The non-commercial energy sources may be divided into two categories for convenience. In the first category fall the new and renewable sources such as solar, wind and geothermal energies, while the second category includes all biomass-based systems like fuel-wood and biogas, which constitute more than 56% of the energy for about 80% of the Indian population. Increased application of renewable energy sources can reduce the consumption of fossil fuels and reduce environmental pollution. Renewable energy systems, therefore, need to be accorded a very high priority in order to provide additional amount of energy on a decentralized basis and reduce air pollution. There is also an urgent need for the in-depth study and development of biomass-based energy sources such as fire-wood plantations, agricultural residues, vegetable oils, liquid wax, hydrocarbon plants, aquatic plants and biological hydrogen.

7.8.4 Conservation

In most of the developing countries, alleviation of poverty is a major social objective. Unless economic development proceeds, it will be impossible to reduce pressures on terrestrial ecosystems in extensive tracts of the world. Societies in their quest for economic development must come to terms with the realities of resource limitation, including the limited carrying capacity of various ecosystems and must take into account the needs of future generations. This is the central message of the **World Conservation Strategy**, which was prepared in 1980 by the International Union for the Conservation of Nature and Natural Resources (IUCN) in cooperation with the United Nations Environment Programme (UNEP) and the World Wildlife Fund (WWF).

The **World Conservation Strategy** has identified four priority requirements for conservation, which are as follows :

(1) The biosphere processes need to be maintained.
(2) The ecosystem resources should be managed and harvested within the limits of their productive capacities and over exploitation should be avoided.
(3) The genetic diversity of wild and cropped biota should be preserved.
(4) Conservation of species and their habitats should be seen as an integral component of the development of natural resources.

Integration of conservation with economic development is primarily a national responsibility and needs to be translated into action at the field level. International assistance should provide support for the national needs of developing countries since the international community shares the global biosphere of the earth.

7.8.5 Control of Thermal Pollution

Power plants, especially the thermal and nuclear ones, constitute the major sources of the thermal pollution of rivers and lakes. Thermal power plants also discharge large quantities of heat into the atmosphere, as do most manufacturing operations and densely populated towns and cities. In thermal and nuclear power plants, the cooling water is usually returned to its source (a lake, a river or a sea) after it has been used for cooling and condensing of exhaust steam from the turbines. The cooling water discharges large amount of heat into the water body. This excess heat has very adverse consequences for the aquatic ecosystem. Rapid changes in temperature produce thermal shock in aquatic life and sometimes, almost immediate death. As water temperature rises, the concentration of dissolved oxygen gets diminished, affecting adversely the aquatic animals. If the water body contains chemical pollutants, extra warmth increases their toxicity to fish. Higher temperature also influences the physical and chemical properties of water.

The problem of thermal pollution can be alleviated by using artificial cooling lakes, ponds and cooling towers. Another method to reduce thermal pollution is to use waste heat for a number of other purposes (like heating a swimming pool or a building), which can simultaneously help to conserve fuel resources. Waste heat can also be used for the desalination of sea water to obtain fresh water.

7.8.6 Recycling and Substitution

Recycling can take the form of re-using mineral land, other materials that have been processed into products and subsequently discarded after use, or recapturing materials used in processing. Recycling of materials can occur at many points in a resource/material flow. The use of scrap iron in iron and steel industry and the recovery and re-use of lead from batteries are obvious examples of recycling at the end-use stage of what otherwise would be waste materials.

Secondary resource recovery is a form of recycling at the processing stage of resource materials flow, by recovering valuable mineral constituents in metallurgical wastes. An excellent example of secondary resource recovery is offered by the bauxite/aluminium industry. Waste dross containing aluminium is generated in the processing of bauxite into aluminium, part of which is recovered which, in turn, results in slag. This slag contains fluxing salts, aluminium metal and aluminium oxide. A hydrometallurgical process has been introduced that is capable of recovering the salt flux and a high proportion of the residual metal.

Another way in which new supplies of a particular mineral may be augmented is by substitution. In some cases, a renewable material can be substituted for the one that is nonrenewable. More often, abundant materials are substituted for the scarce ones in the composition of consumer and industrial products. Substitution can also occur in terms of the design of end products or of the processes of production themselves, away from products and processes that are resource or energy-intensive towards those that are more conservative in terms of resource or energy use.

7.8.7 Development of Eco-friendly Technologies

Industries can reduce their harmful emissions to the environment by :

(1) Adopting processes and materials that generate fewer potentially damaging substances.

(2) Adopting better techniques for recovering substances from liquid and gaseous emissions prior to discharge.

(3) Adopting new means of utilizing and recycling production wastes.

The solution adopted inevitably depends on assessment of the harm likely to be done from an emission or discharge and the benefits the industry confers on the community. If the risk of severe environmental damage is great, stringent control measures will obviously be demanded. Eco-friendly technologies, giving less pollution at less cost, are clearly important for reducing imbalance in the ecosystem and governments have a role in supporting research and development in this area.

7.8.8 Development of Low-waste Technologies

The generally agreed and very broad definition of non-waste or low-waste technology, as determined by the European Economic commission (EEC), is

"the practical application of knowledge, methods and means, so as to provide (within the needs of man) the most rational use of natural resources and energy and to protect the environment".

In essence, a low-waste technology minimizes the creation of harmful wastes. It is a **"preventive"** strategy as distinct from the **"corrective"** approach. Experience has shown that, in many circumstances, it is more efficient and less expensive to incorporate pollution-preventive measures than to implement pollution-corrective techniques at a later stage. Because industrial development will continue for a long time to be an important source of wastes, notwithstanding the new technologies that reduce the amount of waste, it is important to devise methods of reusing the wastes as secondary raw materials. An example of this approach is the reprocessing of blast furnace slag in steel industry to manufacture slag cement and other construction materials.

Industry is critically involved at all stages of the recycling chain. In the first place, product design and composition is an important factor in recycling since most products in current use were not earlier designed with their economic recycling in mind. Secondly, products can also be made with fewer materials, or made to last longer. Thirdly, through the choice of appropriate processes and technologies, waste generation in the manufacturing process can be reduced. In the fourth place, the reclamation industry may be profitable in many cases and, thereby, be responsible for a major part of the recovery of waste materials. Finally, the manufacturing industry may complete the cycle by deciding to use the recovered secondary materials to supplement the virgin materials. In order to encourage and promote recycling of wastes and residues, due consideration must be given to R & D institutions, logistics, markets, technologies and incentives. The keys to successful waste utilization are a beneficial use, an adequate market, an appropriate technology to process the waste under local conditions and an overall enterprise that is socially and economically feasible.

7.8.9 Use of Alternate Resources

Scarce resources can be conserved by substituting other more abundant types of resources. There are many substitutes available for most industrial materials. For example, minerals can be substituted for wood in construction materials, natural rubber can be replaced by synthetic rubber and waste paper or hard woods can be substituted for soft woods in the manufacture of pulp. Decisions about substitution and the use of alternate materials generally depend on economics.

The choice of resources to be used in production of given material is based on the quantity and quality of their long-term supply, the economics of the exploitation of those resources, the energy that must be used to convert the resources to useful forms, the environmental impact of the whole operation and many other social and institutional factors. A substitution strategy, moreover, does not necessarily imply the total substitution of one form of resource by another.

An important aspect of the use of alternate resources is the use of appropriate, simple and low-cost technologies that make good use of indigenous labour and locally available materials. This approach is attractive in developing countries like India having a large pool of labour and high rate of unemployment. The combination of appropriate technology and modern engineering concepts may lead to effective economic and environmental solutions for India.

MODEL QUESTIONS

(Essay/Long Type)

7.1 What are the six classes of the components of the ecosystem? Describe in brief.
7.2 Describe the ecosystem with reference to a pond as an example.
7.3 Write a critical note on the principles of ecobalance.

7.4 Write a short note on biosphere cycles.

7.5 Explain in brief the carbon dioxide cycle.

7.6 Explain in brief the main causes of eco-imbalance.

7.7 What are the adverse effects due to eco-imbalance ?

7.8 List and explain the pragmatic remedies to prevent eco-imbalance.

(Objective/Short Type)

7.9 Give a suitable definition of ecology.

7.10 What is a food chain ?

7.11 What are the decomposers ?

7.12 Define detritus.

7.13 The abiotic components of the ecosystem are :

    (a) Organic compounds   (b) Microconsumers

    (c) Inorganic substances   (d) None of these.

7.14 Macronutrients are :

    (a) Oxygen   (b) Sulphur

    (c) Chlorine   (d) Sodium

7.15 Match the following:

    (A) Ecology   (a) Microbes

    (B) Biotic   (b) Microclimate

    (C) Synecology   (c) Water

    (D) Abiotic   (d) Meteorology

7.16 Match the following:

    (A) Biosphere cycle   (a) Calcium carbonate

    (B) Chlorophylls   (b) Magnesium

    (C) Sedimentary cycle   (c) Biogeochemical cycle

    (D) Oysters   (d) Sulphur cycle

7.17 Write true or false:

An ecosystem may be a single tree.

7.18 Write true or false:

The chemosynthetic bacteria obtain energy from the sun.

7.19 Write true or false:

Fungi, bacteria and animals can be called autotrophs.

7.20 Name the odd ones:

    (a) Carbon dioxide cycle   (b) Sulphur cycle

    (c) Oxygen cycle   (d) Nitrogen cycle

7.21 Fill up the blank:

During photosynthesis, atmospheric CO_2 is incorporated into the production of the carbohydrate which is known as —————.

7.22 Fill up the blanks:

Fats are formed by the elements —————, ————— and oxygen.

Environmental Degradation

8.1   MAN AND ENVIRONMENT

Environment has been defined as *"the sum total of all conditions and influences that affect the development of organisms"*. This is a very comprehensive definition since it stresses the totality of environment, implying that every living organism from the lowest to the highest including human beings has its own environment. Historically, we Indians have been serious minded about our environment and our basic philosophy has been *harmony with nature* in contrast to the Western concept of *conquest of nature*. Unfortunately, we have followed the Western concept during the last 150 years or so, ignoring our own philosophy regarding nature and environment. Since our independence, considerable progress has been made in many fields including agriculture, atomic energy, defence, industry, space technology and information technology. As a result, today we are among the ten most industrialised countries in the world. Associated with any developmental activity, there is always some amount of environmental degradation. Since we did not have a culture of pollution control when India gained independence, we have with us today a huge backlog of more than half a century of environmental pollution and eco-degradation. Time has come now when *sustainability of development* has to enter into our planning process as one of the basic and permanent objectives.

In many ways, India is a distinctive country. India is a developing country among developed nations, but a developed country among the developing ones. Ours is a country where centuries co-exist. For instance, in our transportation system, bullock carts and jet aircrafts are both relevant to our country and so are cowdung cakes and nuclear reactors for our energy requirements. If one visits Andaman and Nicobar Islands or even the interior areas of Bastar district in Chhattisgarh, one finds the tribal living as an inseparable part of nature. They live almost in complete harmony with nature and take from it what they actually require and never more. They generate very little waste, which is biodegradable. Moreover, man is one of the species in an ecosystem, which is self-sustainable and self-generating, with sunlight as the only input from outside. The other extreme is what one finds in our metropolitan cities like Mumbai or Kolkata, where man is the most dominant among all the species. These cities are his world and his environment. He generates huge quanties of waste (much of it-nonbiodegradable) durig production, processing and utilisation of various goods and services. The nonbiodegradable portion of the waste accumulates and degrades the environment. These are the two extremes in India with many situations (showing various degrees of environmental degradation) lying in between. Essentially, the two extremes represent one of *harmony* and the other of *conflict* with nature. However, if we take a look into our past, we find that our history, culture, religion and philosophy have stressed the concept of harmony with nature. In the recent past, however, we ignored this concept and took the Western approach of assault on nature to satisfy our basic needs. As a result, we are in perpetual conflict with nature on account of the indiscriminate exploitation of natural resources. The Western approach stems from a feeling of supremacy over nature. By tradition and culture, we Indians have not been exploiters but utilisers of nature. It was only during the last 150 years or so (and especially since our independence) that we ignored the concept of harmony with nature and, as a result, maximum damage was done to the environment. Concern for the environment, which was a part of our daily life since time immemorial till about 150 years ago, has come to us now via the Western developed countries.

Today the whole world, and especially the developing countries of Asia, Africa and South America, are facing a near-crisis situation on both economic and environmental fronts. Policy makers find it difficult to formulate programmes that would work successfully under the present situation of rapidly growing population on one hand and equally, rapidly diminishing natural resources on the other hand. The environmental degradation invariably weakens the economy which in turn, leads to social disintegration. History is replete with such instances and the remains of past civilisations bear testimony to this. These civilisations were not able to cope with the pressures of degraded environments, and they perished. The link between environmental and socio-economic degradation cannot be overlooked now, particularly because what took hundreds of years in the past is now being telescoped into just a few decades due to tremendous progress in science and technology.

8.2 CAUSES FOR ENVIRONMENTAL DEGRADATION

8.2.1 Economic Development and Environment

In our own country, in the post-independence period, our attitude was dominated by economic development, and we did not have a culture of pollution control. At the time of our independence and in the early nineteen-fifties, we all looked forward to a brave new world of high economic growth through applications of science and technology. In those days, we had unbounded faith in science and technology and the miracles they would work to eradicate poverty, disease and other ills of the society. Since independence, India has made tremendous progress on all fronts. India has also achieved near self-reliance in many core sectors such as iron and steel, crude oil refining and petrochemicals, drugs and pharmaceuticals, fertilizers, engineering goods, food and food processing, building materials and consumer goods. Unfortunately, this industrial transformation of the country was not well-planned from the point of view of environmental conservation and this improperly planned development has resulted in pollution of our air, water and land. The main factor of environmental pollution in the Indian sub-continent (i.e., in India, Pakistan, Bangaladesh, Nepal, Bhutan, Myanmar and Sri Lanka) appears to be over-population that throws heavy demands on natural resources including air, water, soil, flora and fauna resulting in a serious imbalance in the ecosystem. The rich out of their greed and the poor out of their need have been reckless in plundering the earth's assets. Now there is a deep concern about the rapid decline of natural resources, many of these (like fossil fuels) being nonrenewable.

At first sight, industries appear to be the chief villain at the root of environmental degradation. It is a fact, however, that some of our bigger industries produce relatively less process emissions, while sizable emissions come from those industries we often think as minor, e.g., foundries. An extreme example of this was the *Bhopal Gas Tragedy*. In January, 1984, an insecticide plant of Union Carbide leaked a toxic gas, methyl isocyanate, that killed more than 2,000 people within hours and blinded or otherwise sickened thousands. Most of the victims of this biggest industrial accident in the world were the poor people living in the slums adjacent to the plant. It is a sad fact that such potentially hazardous industrial facilities are scattered throughout Indian's crowded cities. Moreover, developing countries like India still remain poorly equipped to protect their environment. The reason is that even if there are appropriate environmental regulations in the developing countries, they are hard to enforce due to lack of trained manpower and other resources.

According to many authorities, the main problem concerning environmental degradation is not "growth" but "unplanned growth". These authorities have formulated a new concept known as "organic growth", as an answer to the problem of environmental degradation. According to the principle of organic growth, the only answer to this universal problem is that governments should make universal decisions above their short-term and narrow national interests. Rich countries should transfer more of their resources to poor nations, and each country should specialise in what it is best equipped to produce.

Development in harmony with the environment should be the spirit of our Five-Year Plans. If we build a large dam, for example, hundreds of thousands of trees are felled, thousands of acres of fertile land is lost, and thousands of poor people are displaced. This is the dilemma of development. No one can deny the need for development; but every effort should be made to minimize its environmental cost. While planning for any developmental project, there should be specific provisions for environmental protection. Attempts should also be made to remove some of the weaknesses existing in the planning system, which include the lack of co-ordinated institutional support, gaps in the necessary information and data required for the assessment of environmental impact of various developmental projects, inadequate availability of trained manpower and lack of strong public interest in environmental issues.

We need to educate the people on the relationship between environment and development. Environment is not merely forests and tigers, endangered species and threatened ecosystems, polluted rivers and overcrowded slums. Environment is the entity on which we all subsist and on which our entire agricultural and industrial developments depend. As a result, development at the cost of environment can take place only upto a point. Beyond that point, it will be like the foolish act of the person who is cutting for fuel the very branch of a tree on which he is sitting. Development without concern for the environment can only be development for a short-term. In the long-term, it will become anti-development which can go on only at the cost of human suffering and increased poverty. Unless corrective measures are taken, we may be rapidly approaching that point.

8.2.2 Environment and the Third World Countries

To understand the nature of environmental problems in India, it is desirable to compare and contrast certain environmental trends in India with those in the advanced countries. Such a comparison is important since many groups in India (including some political parties) dismiss the concern for environment as a petty Western concern and, not very relevant to India. These groups have always argued that too much concern for the environment can only retard economic and industrial development. The major environmental problems in the Western countries are those arising out of waste disposal. These include the problems of air and water pollution, and those of the disposal of highly toxic industrial and nuclear wastes. Though the problem of acid rain has increased in Western countries, many of their cities and rivers look clean. In the Third World countries the problems of air and water pollution are still not the major environmental problems, though they are getting worse day by day as their own industrialization proceeds. In the Third World countries, the major environmental problems are those which arise out of the misuse of their natural resource base, i.e., the misuse of soils, forests and water bodies. These environmental problems are created to a large extent because of the great pressure to produce raw materials for modern industries. The Third World countries provide raw materials to their own domestic industries, and also to the industries of the Western nations; and this creates tremendous pressure on the environment in developing countries. The Third World countries are facing now both an environmental crisis and a developmental crisis; and both of these crises appear to be intensifying and reinforcing each other. On the one hand, there does not seem to be any end to the problems of poverty, inequality and unemployment in the developing countries. On the other hand the problem of environmental degradation remains unsolved and appears to be intensifying. While many environmental problems (especially those related to the pollution of air and water) have become less severe in many of the developed countries due to strict pollution control measures, these problems have only grown and become more severe in many developing countries due to the introduction of highly inappropriate technologies imported form the industrialized countries. Thus, while the process of economic development is tending to solve the problems of environmental degradation in the rich Western countries (due to the introduction of highly capital-intensive pollution control technologies), it is only worsening the environmental problems in poor Third World countries.

8.2.3 Environmental Degradation by the Rich

The food requirements and habit of the rich Western countries have been instrumental in destroying the forests and degrading the land of many poor Third World countries. Despite the worldwide process of decolonization after the Second World War, it is almost certain that much more land is being used today in the developing countries to meet the food requirements of the Western countries than in the years before the Second World War. More than one-fourth of all the Central American forests have been cleared since 1960 for the purpose of cattle ranching and about 90% of the beef produced in these ranches are being exported to the USA, while the domestic consumption of beef in Central America has fallen dramatically during this period. In the US, on the other hand, the beef imported from Central America is mostly used to make pet food and hamburgers. The price of the Central American beef in the US (which is roughly half the price of the beef produced in the US from grass-fed cattle) does not represent its real ecological cost. The reason is that cattle ranching has proved to be the worst form of land use for the fragile soil of Central America on which the tropical forests existed before 1960. It was found that within 5 to 7 years of clearing the forest, the productivity of the soil dropped dramatically and the cattle ranches had to move on to some other areas and clear more forests.

In our own country, the first major attack on the forests of the North-Eastern India came when tea plantations were established. The current over-fishing on India's coast is taking place mainly because of the heavy demand for prawns in the markets of rich Western countries and Japan. This over-fishing often leads to tension and violent encounter between traditional fishermen and the owners of the mechanized trawlers. As a result, India and several other countries in South-East Asia have established regulations to prevent trawler operators from fishing in the first few kilometers from the coast, a zone reserved for the traditional fishermen. Similarly, the export of frog-legs from India to France and other European countries to cater to the palates of rich Western consumers has led to tremendous increase in the population of agricultural pests in the affected areas. To control these pests, farmers have to use large amounts of costly pesticides which, in turn, leads to more environmental pollution.

The pattern of mindless exploitation of natural resources that we see on the global scale simply repeats itself on the national scale. Exactly what the industry of the rich Western countries does to the environment of poor Third World countries, the Indian industry does the same to the environment of the resource-rich and economically poor states of India like Orissa, Jharkhand and Chhattisgarh. Nearly half the industrial output in India comes from biomass-based industries like cotton, textiles, rayon, paper, plywood, rubber, soap, sugar, tobacco, jute, chocolate, tea, coffee, food processing, packaging and vegetable oils. Each of these industries exerts heavy pressure on the cultivated and forest lands of the country. These biomass-based industries need crop lands, forests, irrigation and energy. Thus, we see that one of the main sources of environmental degradation in the world is the heavy demand for natural resources by the high rate of consumption of the rich, whether they are rich nations or rich groups and individuals within the nation itself, and it is mainly their wastes that contribute to the global pollution.

8.2.4 Environmental Degradation by the Poor

It is a curious fact that though the rich are the main contributors to environmental degradation, it is the poor nations as well as groups and individuals within a nation who are affected the most by environmental degradation. As a result, eradication of poverty in a country like India is simply not possible without rational management of our environment and conversely, environmental destruction will only intensify poverty in India. The reason is that the vast majority of rural house holds meet their daily needs through biomass-related products, which are mostly collected freely from the environment. The daily needs of rural India for food, fuel, fodder, fertilizer (organic manure, forest litter and cattle dung), building materials (bamboo poles, thatch, etc.), clothing, herbs, etc., are all biomass products. Although water, which is crucial for survival, is not a biomass product, its availability is closely related to the level of biomass available in the surrounding environment. When the neighbouring forests are destroyed, the local ponds silt up due to soil erosion, the village wells dry up, and the perennial streams are reduced to seasonal ones. The water balance gets totally upset with the destruction of vegetation. This results in greatly increased runoff and floods during the peak rainy season and highly intensified droughts and water scarcity during the lean dry season.

The magnitude of India's dependence on biomass for meeting crucial household needs can be clearly seen by looking at our energy situation. Although we are one of the ten most industrialized countries in the world, it is a fact that more than 50% of the fuel consumption in India is for cooking. In developed countries, on the other hand, cooking consumes less than 10% of the total national fuel consumption. But even more important for India is the fact that more than 90% of the cooking fuel in our country is biomass, i.e., firewood, cattle dung and crop wastes. Many urban households are heavily dependent on firewood for cooking. Biomass resources not only meet crucial household needs, but also provide a wide range of raw materials for traditional occupations and crafts. Thus, biomass resources are a major source of employment. Firewood and cattle dung are important sources of fuel for potters, bullock carts are made from wood, bamboo is a vital raw material for basket weavers and so on. Traditional occupations and crafts are now threatened not only by the introduction of modern products, but also by the acute shortage of biomass-based raw materials. Fodder is another vital resource that is in acute shortage. With only 2.45% of the world's land mass, India supports about 15% of its cows, 52% of its buffaloes and 15% of its goats. These animals play an extremely important role in the integrated system of agriculture practiced by farmers in India. Acute shortage of fodder means that poor landless household and marginal farmers do not get much benefit from milk co-operative societies and animal improvement schemes introduced in the region. Despite such a heavy reliance on biomass resources for bare survival, nature in India has steadily undergone a major transformation. At present, there are two major pressures operating on the country's natural resources. The first and the well-publicised pressure is the one generated by the rapid growth of our population, which puts heavy demand on our biomass resources. The second major pressure on our natural resources comes from modernization, industrialization and the growth of cash economy with profit motives.

8.2.5 Environmental Degradation by Transformation of Nature

Modernisation affects nature in two ways. In the first place, it affects the environment in its search for cheap biomass-based raw materials and cheap methods for waste disposal. Unless there are strict environmental laws, which are strictly implemented, there is hardly any attempt by industries to internalize environmental costs. Both private and public-sector industries prefer to pass on the environmental costs to the society in order to keep their operating costs low and make more profits. Even, State Governments give away large tracts of forests at low cost to industrialists in order to get a few more factories established in the state. In addition to the destruction of the environment, there is another way in which modernization affects the environment, viz., by steadily transforming the very character of nature. In physical terms, modernization has the tendency to remove the **diversity** in nature and replace it by high-yielding monocultures. When this transformation takes place, the ecological role of the original nature (having more diversity) is also degraded. In social terms, the transformation from diversity to monocultures is generally away from the nature supporting household and community needs towards a nature that is geared to meet urban and industrial needs. Some examples of such a transformation of nature are the pine forests in place of the old oak forests in the Himalayas, the teak forests in place of the sal forests in the Chhotanagpur Plateau, eucalyptus plantations in place of natural forests in the Western Ghats and oil palm plantations in place of tropical forests in the Andaman and Nicobar Islands. Both of these phenomena (destruction of the original nature for modernization and the creation of a new, commercially-oriented nature) have been taking place simultaneously and on a massive scale, in the Indian environment. The effect of this massive environmental change has been disastrous to the people. In a country like India, with an extremely high level of poverty and a fairly high level of population density there is hardly any ecological space left in the physical environment which is not occupied by one human group or another for its sustenance. In the name of economic development, therefore, if any human activity (e.g., the construction of a large dam) results in the destruction of an ecological space (or its transformation which benefits mostly the more powerful groups in the society), then those who were earlier dependent on that space for sustenance will suffer. Development in some cases leads to displacement of a large number of people and raises the question of social injustice. The planting of eucalyptus trees on farmers' fields (and even on the barren fields) is an excellent example of the adverse consequences to the poor people living in that area. When a patch of barren land is planted with eucalyptus, the weeds which were growing on the land earlier are no longer available to the poor households, thus intensifying their fuel crises.

Thus we see that in India (and also in many other developing countries), there is today a growing conflict over the use of natural resources in general, and over the use of biomass in particular, between the two sectors of country's economy. On the one hand, there is the powerful modern sector (i.e., the cash economy). On the other hand, we have the traditional sector of biomass-based subsistence economy. The destruction of the environment and its transformation in India is also affecting, on a daily basis, the artisans, nomadic people, the tribals, fishers folk and women from landless or marginal households. These groups add up to more than half of the country's entire rural population. In rich Western countries, the question of environmental degradation is related to the quality of life. In India and many developing countries, preventing environmental degradation is a matter of survival for the poorer sections of the society.

8.3 EFFECTS OF ENVIRONMENTAL DEGRADATION

8.3.1 Destruction of Forests

One of the major effects of environmental degradation due to industrialization is the large-scale destruction of our forests, and this has a major impact on the productivity of our crop lands. The reduction of crop-land productivity due to the destruction of forests occurs in two ways. In the first place, when forests are destroyed, there is a many-fold increases in the rate of soil erosion. The result is that the soil literally gets washed leading to an intensified cycle of floods and drought. Secondly, the destruction of forests leads to the shortage of firewood and this has an adverse impact on the productivity of crop lands. The reason is that when firewood becomes scarce, people begin to use cowdung and crop wastes as fuel for cooking and other purposes. As a result, every part of crop plants gets used (as fodder or fuel) and nothing goes back to the soil to enrich it. Over a period of time, this constant drain of nutrients from the soil adversely affects it productivity. The more intensive the agriculture (due to the use of high-yielding varities), without any manure and crop wastes going back into the soil, the faster is the nutrient drain. The district of Ludhiana in Punjab, for example, has the highest yield per hectare of many cereals today; but Ludhiana has also the highest deficiency of many micronutrients in its soil. In Punjab, many farmers have already started using zinc routinely as a fertilizer. If the drain of micronutrients from the soil continues, the farmers will soon be using sulphur, manganese and iron as fertilizers. The micronutrient fertilizer industry is likely to become a boom industry in the near future.

8.3.2 Degradation of Grazing Lands

If the existing crop lands and water resources for irrigation are not used efficiently, then due to a rising population, the demand for using marginal lands for agricultural purpose will grow. As large parts of our country have excellent soils and the only shortage is that of water, government programmes have also promoted the cultivation of marginal lands (especially through the spread of irrigation schemes). This policy of expansion of cropped areas has already resulted in enormous ecological damage. For instance, crop lands have expanded and have encroached on grazing lands and this has resulted in overstocking of grazing lands, thus destroying their productivity and impoverishing the graziers in the process. Faced with the decreasing trend of grazing lands, the graziers have followed two strategies. As the grazing land becomes scarce and the environment becomes more and more hostile, the graziers try to get rid of the more vulnerable cattle and start keeping goats as the first strategy. The number of goats in Rajasthan, for instance, has increased dramatically and much faster than the number of any other livestock. Although goats are highly destructive of the environment (by destroying the vegetation cover), they are better suited to the hostile environment that the human beings are creating in Rajasthan and other arid parts of the country like Gujarat and Maharashtra. For the grazier, it makes economic sense to reduce his risk by keeping more goats during a period of drought and this practice is quite common in Rajasthan, Gujarat and Maharashtra. There is yet another way in which the graziers try to solve their problems, when faced with decreasing grazing lands or droughts. They begin to use forests as grazing lands. In fact, India's forests are among the most heavily grazed forests in the world. As forests are now disappearing in Rajasthan and Gujarat, graziers and nomads from these states now enter Madhya Pradesh in large numbers since Madhya Pradesh is still a heavily forested state. Faced with all these problems, the experts sit in grand isolation. Foresters have no interest in fuelwood or crop lands. Agricultural experts have no interest in animals or grazing lands. Experts in animal husbandry never advise the foresters to make arrangements for fodder. Such isolated thinking plays havoc on the environment and finally, nature turns against itself. Once the forests are denuded, the solar energy and high temperature begin to bring high soil erosion, ecological destruction and social poverty. Given the climatic conditions of India and other developing countries of South-East Asia, unchecked soil erosion has a very large negative impact on soil productivity in these countries. In order to solve all these problems of environmental degradation, there must be more holistic thinking regarding the management of land and water resources, and this will not be easy unless a determined effort is made.

8.3.3 Adverse Effects of Air Pollution

Air and water are among the most important constituents of the life support system. It is well known that one may live for several weeks without food, and for several days without water. Without air, however, one cannot live for more than a few minutes since a lack of oxygen leads to permanent damage and death of the brain cells. Air pollution results from industrial activities and also from dense human settlements. Gaseous emissions are discharged into the atmosphere by industries, thermal power plants, automobiles and other vehicles, trains, airplanes, domestic cooking and heating, and so on. Air pollution is also related to the meteorology (i.e., weather and climatic factors) and anthropogenic factors of a particular locality. In India, for instance, even in the absence of heavy industrialization, the domestic energy systems and transportation contribute heavily to air pollution. Although there are many parameters for judging the quality of ambient air, the National Environmental Engineering Research Institute (NEERI) at Nagpur has recommended three major parameters, which give a fair idea of the extent of air pollution. These three parameters are: (1) oxides of sulphur (SO_2 and SO_3, known collectively as SO_x), (2) oxides of nitrogen (mainly N_2O, NO and NO_2, known collectively as NO_x) and (3) suspended particulate matters (SPM). There are numerous adverse effects of air pollution on human beings, animals, plants, buildings and materials like cloth, metals, paints and paper.

8.3.3.1 Oxides of Carbon

There are two common oxides of carbon, CO and CO_2. Most gaseous pollutants only irritate the respiratory system, but the action of carbon monoxide on the human body is unique. On inhalation, carbon monoxide passes through the lungs into the blood stream, where it combines with the haemoglobin (Hb) of the red blood cells to form a complex known as carboxyhaemoglobin (COHB):

$$Hb + O_2 \longrightarrow HbO_2$$
$$HbO_2 + CO \rightleftharpoons COHb + O_2.$$

When the carbon monoxide (CO) present in the polluted air binds with the haemoglobin, it interferes with the haemoglobin's natural function of carrying oxygen to the cells in different parts of the body. The normal level of COHb in the human blood cells is about 0.5%. When the COHb level reaches above 5% (due to carbon monoxide poisoning), the cardiac and pulmonary functions are affected, and severe carbon monoxide poisoning (i.e., exposure to high level of CO for longer periods) may lead to coma and death. Unlike carbon monoxide, carbon dioxide is not toxic; so it is not harmful directly to human health. However, if the concentration of CO_2 increases in the atmosphere, it has a long-term effect known as the "*Greenhouse Effect*". The CO_2 molecules absorb the heat radiated by the earth (just like the glass walls in a greenhouse) and do not allow the heat energy to escape into the outer space. As a result, the temperature of the earth's atmosphere increases. This phenomenon is known as the "*Global Warming*", and it has many long-term adverse effects like melting of polar ice caps, adverse changes in the climate and rainfall patterns, reduction in agricultural productivity, rising of sea level, submerging of coastal areas, and reduced fish yield from the oceans.

8.3.3.2 Oxides of Sulphur

The two oxides of sulphur are sulphur dioxide (SO_2) and sulphur trioxide (SO_3), which are collectively known as SO_x, where x=2 or x=3. When any fuel containing sulphur (such as coal) burns, the first product of sulphur is SO_2;

$$S + O_2 \longrightarrow SO_2$$

In the presence of solar radiation, the SO_2 is oxidised to SO_3;

$$3SO_2 + O_2 \xrightarrow{\text{sunlight}} 2SO_3$$

When SO_3 comes in contact with water vapour present in the atmosphere, sulphuric acid is produced;

$$SO_3 + H_2O \longrightarrow H_2SO_4$$

This sulphuric acid returns to earth with the rainwater or snow. The phenomenon is known as "**Acid Rain**", which has many harmful environmental effects. Acid rain destroys vegetation and forests, increases soil acidity (which results in reduced soil productivity), increases acidity in rivers and lakes (which leads to destruction of aquatic life) and slowly destroys buildings and ancient monuments. The action of H_2SO_4 on marble (lime stone) occurs as follows;

$$CaCO_3 + H_2SO_4 \longrightarrow CaSO_4 + H_2O + CO_2$$

This attack on marble is known as "**Stone Cancer**" or "**Stone Leprosy**". As a result of stone cancer, marble becomes brittle and discoloured, and slowly crumbles. It was to save the Taj Mahal from acid rain that the Supreme Court of India gave orders to remove foundries out of Agra. Acid rain also corrodes metals, attacks fabrics and washes out alkaline salts from the soil. In addition to the formation of H_2SO_4 and the resulting acid rain,

SO_2 and SO_3 have other harmful effects since they are both toxic to human beings and animals. The main effect of these gases is on the respiratory system. In low concentration in the ambient air, they cause eye irritation and discomfort. Inhalation of SO_2 in higher concentration causes the symptoms of bronchitis, emphysema and other lung diseases including lung cancer. Exposure to very high concentrations of SO_2 for longer period may lead to acute sickness and death.

8.3.3.3 Oxides of Nitrogen

There are several oxides of nitrogen (such as N_2O, NO, NO_2, N_2O_3 and N_2O_5), which are collectively known as NO_x. These are largely emitted by motor vehicles and thermal power plants using fossil fuels like coal and oil. Among the various oxides of nitrogen, N_2O is present in the air at trace concentration level of about 0.25 ppm (parts per million). It is not a product of combustion. The main product of combustion of nitrogen-containing fuels is NO, which is oxidized to NO_2 by O_2 or O_3;

$$2NO + O_2 \longrightarrow 2NO_2$$
$$2NO + 3O_2 + N_2 \longrightarrow 4NO_2$$
$$NO + O_3 \longrightarrow NO_2 + O_2$$

Nitrogen dioxide (NO_2) is a strong absorber of ultraviolet light in the solar radiation, so it initiates photochemical reactions in the troposphere and produces "smog" (i.e., smoke plus fog), which obstructs the sunlight. N_2O_3 and N_2O_5, on the other hand, react with water vapour present in the atmosphere to form nitric acid (HNO_3), which gives rise to acid rain when it returns to earth with the rainfall or snowfall.

Nitrogen dioxide (NO_2) is more toxic to human beings compared to nitric oxide (NO). NO_2 has an irritating effect on the mucous membrane. In case of vegetation, higher concentration of NO_2 damages the leaves of some plants and retards the process of photosynthesis. Higher concentration of NO_2 in the atmosphere may lead to bronchitis and other lung diseases. NO, on the other hand, gets attached to the haemoglobin and reduces its oxygen transport efficiency.

8.3.3.4 Suspended Particulate Matters

Small solid particles and liquid droplets are collectively known as particulates. They are present in the atmosphere and often pose a serious air pollution problem. The size of these particulates varies from 0.0002μ to $500\,\mu$ ($1\mu = 10^{-6}$m). Their lifetime in the atmosphere ranges from several seconds to several months (depending on their size). Suspended particulate matters (SPM) include organic matters, nitrogen compounds, sulphur compounds, several metals (including the toxic ones), radioactive substances and polycyclic aromatic hydrocarbons (PAH). Suspended particulate matters have numerous harmful effects on human health. Workers exposed to asbestos dust can develop a type of cancer known as mesothelioma, which occurs in the tissue lining of the abdomen. These workers also have a greatly-elevated risk of lung cancer. Similarly, the inhalation of dust containing free silica (SiO_2) causes a disease known as silicosis. "Black lung" is a common disease found among coal miners, whereas a disease known as "white lung" is quite common among textile workers. SPM also include dust, soot, fumes and mist, which are potentially harmful to various materials. Particulates accelerate corrosion of metals, which is found to be maximum in industrial areas. Particulates also cause damage to buildings, paints and furnitures.

8.3.4 Adverse Effects of Water Pollution

8.3.4.1 Pollution of Water

Water is indeed a wonderful chemical compound with the unique property of dissolving and carrying in suspension a huge variety of other chemicals. This unique property of water, however, has an undesirable effect, viz., that water gets contaminated very easily. Pollution of water makes it unsuitable for its normal uses in agriculture, public water supply and industry. Polluted water is harmful also to aquatic life. Water may be polluted due to the following major pollutants:

(1) Inorganic pollutants,
(2) Organic pollutants,
(3) Sediments,
(4) Thermal pollutants (heat) and
(5) Radioactive substances.

8.3.4.2 Inorganic Pollutants

The inorganic pollutants of water consist of mineral acids, inorganic salts, metal compounds, finely-divided metals, organometallic compounds, trace elements and polyphosphates. Coal mines discharge considerable amount of sulphuric acid and ferric hydroxide into local streams through seepage. Sulphuric acid (H_2SO_4) and ferric hydroxide ($Fe(OH)_3$), which are added to water bodies through seepage of acid mine discharge, destroy aquatic life. Organometallic compounds have an adverse impact on the toxicity of metals in aquatic ecosystems and they also lead to unchecked growth of algae in water bodies. A large number of trace elements are found in polluted water. Out of these, the most hazardous trace elements are the heavy metals like Hg, Cd, Pb, As, Se and Sb. Mercury (Hg) has been known to be toxic to human beings and animals for centuries. Mercury poisoning leads to permanent impairment of the nervous system and in extreme cases, to death. It causes sensory loss in limbs, impaired vision and hearing, change in personality and loss of intellectual capacity. Severe exposure to mercury vapour produces tightness and pain in the chest, difficulty in breathing and coughing, followed by sleeplessness and headache. Milder mercury poisoning results in muscular tremors, while severe exposure leads to sour mouth, severe irritation of the pulmonary tract, inflammation of kidneys and malfunction of the brain. Discharge of cadmium (Cd) into water bodies occurs from the electroplating industry, which accounts for about 50% of Cd in water. Other sources of Cd are nickel-cadmium battery industries, fungicides and fertilizers. Cadmium builds up in the human body over a long period of time. It accumulates in the kidneys, liver, lungs and pancreas. Chronic cadmium poisoning leads to proteinurea and causes kidney stones. Low level of cadmium poisoning results in high blood pressure and heart disease along with symptoms like rheumatism, neuritis and bone pain. Acute cadmium poisoning leads to vomiting, abdominal cramps and general weakness. The major biochemical effect of lead (Pb) is its interference with hemaesynthesis which leads to hematological damage. Inorganic Pb^{+2} is a metabolic poison. Younger children are particularly susceptible to lead poisoning and when exposed to lead, they suffer from mental retardation and semi-permanent brain damage. Lead is very toxic also to animals and plants. Tetraethyl lead (used as an antiknocking additive in motor vehicle fuels) is even more toxic than inorganic lead. Initially, it causes excitement, depression and irritability. Children exposed to tetraethyl lead show progressive deterioration of mental ability. Severe exposure to it causes clumsiness, change in mental attitude, poor memory, restlessness, severe depression of the nervous system and finally, death.

8.3.4.3 *Organic Pollutants*

Organic pollutants include oil, pesticides, herbicides, detergents, synthetic organic compounds, and fertilizers. Organic pollutants enter the water bodies (like rivers, streams and lakes) through the discharge of domestic sewage, industrial effluents from food procession plants and paper mills, waste from slaughter houses, and run-off from agricultural land. When organic pollutants enter the water bodies, the water system undergoes degradation due to bacterial activity, which consumes the dissolved oxygen (DO). Dissolved oxygen is an essential requirement for the existence of aquatic life. Decrease in DO level in water bodies, which is an indication of their contamination with organic pollutants, adversely affects the population of aquatic plants and animals. Deoxygenation of water occurs when bacteria decompose organic pollutants and convert their carbon to carbon dioxide:

$$C + O_2 \longrightarrow CO_2$$

The CO_2 produced by this process dissolves in water to form carbonic acid ($H_2 CO_3$) which increases the acidity of water and adversely affects aquatic plant and animal life.

The sources of oil pollution are oil spills from cargo oil tankers on the sea, losses of oil during off-shore oil exploration and production and leakage from oil storage tanks and pipelines. A layer of oil on the surface of water reduces the transmission of sunlight through the surface. This reduces the rate of photosynthesis by marine plants which, in turn, reduces the amount of oxygen dissolved in water and adversely affects aquatic life. Pesticides are organic chemicals which include herbicides (used to kill weeds and undesirable vegetation), fungicides (used to eliminate fungi or moulds and control plant diseases), and insecticides (used to kill insects). Pesticides have played a very important role in eradicating diseases such as malaria and typhus, and also in increasing the production of food. Among the pesticides, the biological action of DDT has been studied in great detail. DDT is a chlorinated hydrocarbon, which has been widely used in the past for the eradication of pests, insects and insect vectors (such as mosquitoes). DDT is a fairly stable and persistent chemical. Once introduced into the environment, DDT keeps circulating for many years. DDT dissolves in lipids, accumulates in fats and targets the central nervous system. DDT is also believed to be a carcinogen (a cancer producing agent) in man. It can cause tumour in the liver. DDT has also been found to adversely affect the human reproductive system by killing sperms and lowering the fertility. Premature births, aborted foetuses, lower birth weights and brain damages have also been associated with DDT. Detergents are organic chemicals used as cleaning agents. Detergents are also responsible for water pollution. They form a layer of foam on the surface of water and make its appearance unaesthetic. Moreover, the phosphates provided by detergents act as nutrients for aquatic plants and lead to the phenomenon of eutrophication (i.e., excessive growth of aquatic plants like moss) in water bodies.

8.3.4.4 *Sediments*

Soil erosion by natural processes (or increased soil erosion due to deforestation and other human activities) gives rise to sediments in water bodies. Sediments are among the major pollutants of surface water (river, streams, lakes and ponds). In addition to deforestation, soil erosion is also increased due to agricultural development, construction activities and strip mining. Sediments are the major sources of organic and inorganic matters in water bodies. The level of organic matter in sediments is usually higher than in soils. Bottom sediments have the capacity to exchange cations (positive ions) with the surrounding water medium. Sediments are, therefore, an important source of trace elements (such as Cu, Mo, Ni, Co, Mn, Cr, Pb, Cd, Hg, As, Sb, etc.) in water.

8.3.4.5 *Thermal Pollutants (heat)*

Most thermal power plants utilise coal as the fuel while some of them use oil. Nuclear power plants, on the other hand, utilise enriched uranium as the source of energy. All these power plants are associated with the problem of thermal pollution. In case of thermal and nuclear power plants, thermal energy released from the combustion of fossil fuels (coal or oil) or the fission of uranium is utilised to produce steam. This steam is utilised to produce electricity by running steam turbines, which are coupled to generators. As a consequence of the second law of thermodynamics, only a fraction of the thermal energy (i.e., heat) is utilised to do useful work (i.e., to produce electricity) at the thermal or nuclear power plants. The rest of the heat is wasted and discharged into the environment. In case of coal-fired power plants, the maximum efficiency is 40%. So, 60% of the heat produced by burning coal is discharged into the atmosphere and nearby water bodies. The condenser coils of the power plant are usually cooled with water from a nearby lake or river. This cooling water, after absorbing heat from the steam, is discharged back into the lake or river concerned. The same is true for nuclear power plants. This process of discharging large amounts of heat into nearby water bodies raises the temperature of water. Such warm water has very adverse effects on aquatic flora and fauna. In the first place, when heat is added to water, it reduces the amount of dissolved oxygen in it since the solubility of oxygen in water decreases with temperature. Finally, excess heat added to water bodies results in killing fish and other aquatic animals since they cannot survive at higher temperatures.

8.3.4.6 *Radioactive Substances*

In addition to vast amounts of heat, the nuclear power plants also discharge traces of radioactive substances into nearby water bodies, thus causing harmful effects on aquatic life. Nuclear power plants, however, are not the only source of radioactive pollution of the environment. Other major human activities responsible for the radioactive pollution of air, water and soil are :

(1) Testing of nuclear weapons,
(2) Nuclear power plants,
(3) Use of radioactive isotopes in medicine, agriculture, industry and research laboratories and
(4) Mining and processing of uranium and other radioactive ores.

In the processing of uranium are, one has to start with a large amount of uranium ore. Out of that, only a very small amount of ore is rich in uranium, which is used for uranium extraction. As a result, vast quantities of uranium ores end up as "uranium tailing" and pollute the environment.

Although testing of nuclear weapons in the atmosphere and oceans has been banned by international treaties, their testing by underground detonations gives rise to radioactive pollution of the sub-surface soil and rocks. The radioactive substances released by underground nuclear explosions may contaminate the ground water in the surrounding areas and, depending on the geology of the area, may even leak to the surface soil and the atmosphere.

Nuclear explosions release many different types of radioactive isotopes. Most of them have a short half-life, so they are transformed to harmless, non-radioactive substances in a fairly short time. Some radioactive isotopes such as Sr^{90} (a radioactive isotope of strontium with atomic weight=90) have a very long half-life. If such radioactive isotopes are released into the ground water, soil or air, they remain there for hundreds of years, and have highly adverse effects on human beings, animals and plants. Such long-lived radioactive isotopes include K^{40}, Mn^{54}, Co^{60}, Kr^{85}, Sr^{98}, I^{131}, Cs^{137}, Ba^{140}, Ra^{225}, U^{238} and Pu^{289}. In the human body, these radioactive substances interfere with blood cell formation and lead to serious anaemia. Some of the long-lived radioactive substances accumulate in the body and they may produce cancer in the organ in which they accumulate.

Nuclear power plants generate the following pollutants:

(1) Fission products (most of which are radioactive),
(2) Liquid and gaseous wastes from fuel elements (some of these being radioactive),
(3) Liquid wastes (with low level of radioactivity) and
(4) Vast amount of heat energy.

When the radioactive wastes from nuclear power plants escape into the ambient air of nearby water bodies, they pose long-term hazards to plants, animals and human beings. One of the most serious hazards of radioactive pollution is the genetic damage caused by nuclear radiations.

8.3.5 Adverse Effects of Soil Pollution

8.3.5.1 Soil and Its Pollution

Soil, like a film, envelopes the surface of our planet. Soil acts as the boundary between the biosphere and the lithosphere. Because of its peculiar structure and composition, soil must be treated as a complicated biogeochemical system.

Soil is made up initially of the rock that has been substantially transformed by the actions of various climatic factors (like rainfall and wind), vegetation and various micro-organisms. The influence of organisms on the rock depends on the climatic conditions, especially on the hydrothermal regime. The process of soil formation depends not only on the rock, but also on the plants, animal organisms, climate, and age of the land mass. Considering these factors, soil may be defined as a natural formation resulting from the transformation of the surface rock by the combined action of the climate, plant and animal life, and ageing.

A very important property of the soil is its fertility. This is the property that has been investigated most thoroughly by soil scientists so that the soil can be used most rationally for agriculture and forestry. Soil can also be a source of useful minerals and fossil fuels. In addition, soil is also used in building construction, land improvement and road construction. The most important property of soil is its absorbing quality. Soil absorbs waste products of plant and animal kingdom, industrial and municipal wastes, and toxic gases; but when the amount of these wastes is too large, soil gets polluted and loses its fertility and other important qualities. Some of the major pollutants of soil are :

(1) Industrial and municipal solid wastes.
(2) Pesticides.
(3) Plastics.
(4) Sewage irrigation.
(5) Electronic waste.

8.3.5.2 *Industrial and Municipal Solid Wastes*

In recent years, it has become very clear that the natural sources of the world are indeed finite. In order to maintain a stable economic growth and a high standard of living in the future, it is very necessary that we use the available resources very carefully, generate the least possible amount of waste and develop suitable technologies for the recycling of wastes and residues, thereby saving both resources and energy. Waste has always been a part of human activity since time immemorial. The baking of bricks for constructing the Great Wall of China saw the destruction of forests all around, thus damaging the environment and impoverishing the country. In earlier times, however, populations was less, needs were few, resources were abundant and the generation of waste was such that it got naturally recycled since it was mostly biodegradable. After the Industrial Revolution and especially during the last few decades, the resources have been used recklessly and there has been generation of diverse types of wastes, which are often both non-biodegradable and hazardous. Since an ecosystem cannot absorb them in the natural course, the piling of such industrial and municipal wastes poses a difficult environmental problem. At the global level, these wastes are likely to affect even the stability of the biosphere.

Industrial and municipal waste (including bio-medical waste) can be a big health hazard in some situations and must be disposed off properly to prevent such situations. At present, most of the municipal solid waste is disposed off for no other use than sanitary land-fill. As a result, new land-fill sites have to be found every year and thus, useful land is wasted. In case of bio-medical waste (i.e., hospital waste), however, a part of it is infectious or hazardous; so it can not be used for land-fill along with the municipal waste. Infectious bio-medical waste has to be burnt in an incinerator. Bio-medical waste contains disposable syringes and other plastic items and contributes to the problem of air pollution.

Fly-ash from coal-fired thermal power plants is a major industrial waste product. Coal is the cheapest fuel in India. As a result, there is a large number of coal-based thermal power plants in our country, and many more such plants are being planned for the future. It requires about 55 tonnes of coal to produce one kilowatt of electricity, and fly-ash constitutes roughly 15-20% of the total coal consumed. So, millions of tonnes of fly-ash are produced in our country each year. Only a small fraction of this huge quantity of fly-ash is utilized to produce cement, bricks or other building materials, while the rest pollutes the soil and water.

Industrial and municipal wastes, if not recycled or properly disposed off, can be potential environmental hazards by spreading diseases and leaching dangerous chemicals into air, water and soil. The world is witnessing now a major crisis in resource availability. The developed countries want to conserve their own resources as much as they can; so they import resources from developing countries (e.g., the US importing oil from the Middle-East countries) in return for some hard currency. In this game, the developing countries are the major sufferers. It is necessary, therefore, that India and other developing countries should recycle their industrial, municipal and agricultural wastes to conserve their resources and reduce environmental degradation.

8.3.5.3 Pesticides

Pesticides are chemicals used to control the harmful organisms in the daily life of modern man. In India, the per capita consumption of pesticides is roughly 0.5 kg. There are three major problems that threaten to limit the continued use of pesticides :

(1) Some pest organisms develop resistance to the pesticide currently in use; so a more potent pesticide has to be developed, which may be even more hazardous.
(2) Some pesticides have detrimental effects on organisms other than the target pests. They may adversely affect the soil, flora and fauna.
(3) Some pesticides are not readily biodegradable, so they tend to persist for years in the environment and pollute soil and water.

These problems require attention of research workers, public health authorities and governments because pesticides are hazardous not only to human and animal health, but also to the economy as a whole. When pesticides are applied to soil (to control the agricultural pests, for example), they act in the following ways :

(1) The chemicals may be absorbed by the soil and pollute it.

(2) They may move downwards through the soil in liquid or solution form by leaching and pollute the ground water.

(3) They may vaporise and cause air pollution by entering into the atmosphere.

(4) They may undergo chemical reactions within or on the surface of the soil.

(5) They may be broken down by soil organisms.

Thus we see that when pesticides are applied to the soil, they lead to soil, air and water pollution. Due to wide-spread use of pesticides and other chemicals, the food consumed by man today is contaminated with hazardous chemicals. These chemicals are stored in the fatty tissues of the human body. It has been estimated that about 5-27 ppm (parts per million) of DDT is found in the tissues of persons living in the countries where DDT is widely used or has been used. The accumulation of hazardous chemicals in the human body has been found to be harmful for health since it may lead to dangerous diseases like hypertension, cancer and sterility. Some chemicals damage the liver and kidneys. The pesticides reach the human body indirectly through vegetables, fruits, eggs, milk and cereals. People have died after eating foods contaminated accidentally with pesticides.

8.3.5.4 Plastics

In the last few decades, the use of plastics has increased all over the world at a very fast rate because of their low cost and many desirable properties. However, the main objection to the large scale use of plastics is that they are non-biodegradable, so they may persist for centuries and pollute the soil. Polythene bags, for example, clog the drains and cause flash floods and water logging in towns and cities after heavy rains. Discarded polythene bags also choke ponds, streams and rivers. When plastics enter into the soil, they have adverse effects on soil fertility and its flora and fauna. Now-a-days, the most popular plastic is the polyvinylchloride (PVC). When any food material or blood is stored in a PVC container, the soluble chemicals in the plastic are gradually dissolved into the food or blood, and they have very adverse effects on human health. That is why, the use of coloured polythene bags for carrying food items has been banned in many states in India, and the Himachal Pradesh Government has imposed a total ban on the use of polythene bags. Harmful substances in the PVC may cause skin diseases and cancer. In case of animals, PVC has been found to harm their respiratory system and destroy their fertility. When the harmful chemicals of PVC mix with water, the polluted water can cause paralysis and damage the bones. The US and some other developed countries have already banned the use of PVC in food containers. It is high time for India and other developing countries to ban the use of PVC in food containers, water pipes and medicine containers to save their citizens from a number of health problems.

8.3.5.5 Sewage Irrigation

Sewage is the liquid waste varying considerably in strength and composition from town to town due to marked differences in the living habits of different communities. Domestic sewage is mainly the discharge of dirty water from bathrooms and kitchens. In addition to organic matter, the sewage also contains living matters like bacteria, fungi and protozoa. Domestic sewage is an excellent medium for the growth of micro-organisms. Although synthetic detergents are detrimental to plant growth at higher concentrations, domestic sewage is relatively free from substances which can prove toxic to plants. However, if industrial effluents are also mixed with the domestic sewage, many toxic substances (depending upon the type of industry) may creep into the sewage.

The use of sewage in farming seems to have been quite popular in the past. In the middle of the nineteenth century, the great chemist Liebig drew attention to the fertilizing value of sewage and even recommended land treatment as the best method for sewage disposal. The development of sewage irrigation can be justified from the view-point of agricultural necessity and economy; but care should be taken to minimize its environmental impact. Sewage irrigation may pollute the soil since many substances contained in the sewage that are beneficial to plants at low levels can be toxic at higher concentration. The sewage may contain organic matter, nitrogen, phosphorous, sulphur and many heavy metals like Cd, Cr, Cu, Hg, Pb, Ni and Zn. Heavy metals like cadmium, zinc and lead may not be detrimental to crops at lower concentration, but they may enter into the food chain and lead to toxic effects on human beings and animals. Heavy metals usually accumulate in the upper soil.

When sewage is applied to the soil, it tends to affect the status of oxygen in the soil, mainly in two ways :

(1) The oxygen in the soil pores is directly consumed by the soil micro-organisms for the decomposition of organic matter contained in the sewage.

(2) Indirectly, the fine organic matter contained in the sewage in semi-colloidal state gets deposited in the voids of the soil. This adversely affects the porosity of the soil and reduces its capacity to exchange air, which leads to reduction in the oxygen status of the soil.

Long-term use of untreated raw sewage at excessive rates of application may adversely affect the soil-air relationship. Ultimately, the soil is rendered unfit for growing the crops economically. Such a condition is often referred to as *"sewage sickness"* of the soil. Nitrogen is another factor that has to be considered carefully when sewage is used for irrigation. Either too much or too little nitrogen adversely affect the growth of crops. Moreover, excessive nitrogen in the form of nitrate can accumulate in the ground water as a pollutant since nitrates are highly soluble and mobile. Many pollutants (such as heavy metals) may accumulate in crops in sufficient concentration to pose a serious hazard to the health of human beings and animals without any visible effect on the growth of crops. The tolerances of plants to the level of these pollutants in the soils vary widely with various plant species as well as with the properties of the soil. Uptake of the soil pollutants depends on both plant species and plant organs. For example, the absorption and accumulation of mercury in the edible parts of crops in the decreasing order are: (1) rice (2) cabbage (3) turnip (4) maize (5) sorghum and (6) wheat. In case of cadmium, however, the crops in decreasing order of accumulation in edible parts are: (1) turnip (2) cabbage (3) wheat and (4) rice. Fortunately, most of the fruits, seeds and grains have a lower concentration of heavy metals than leaves and roots. It is evident, therefore, that in order to alleviate the problem of soil pollution due to sewage irrigation, farmers must be advised to select crops with lower rates of uptake of the pollutants concerned.

8.3.5.6 *Electronic Waste*

It has been estimated that more than one billion PCs (personal computers) have already been sold globally and there has also been a spurt in the sales of other electronic items like refrigerators, air conditioners, cellular phones and personal stereos. As a result, the quantum of electronic waste (discarded electronic items or *"e-waste"*) is increasing day by day. Studies conducted in Europe reveal that the volume of electronic waste is increasing at the rate of 3% to 5% per year, which is almost three times more than the rate of growth of the municipal wastes. It is not only quantity of the e-waste that is alarming, but also the toxic substances contained in them, which pose serious occupational and environmental hazards. The electronic wastes contain more than 1,000 different substances, many of which are toxic. Recent studies indicate that e-waste is a complex mixture of a number of toxic substances like lead and cadmium in computer and batteries, and polychlorinated biphenyls (PCBs) in older capacitors and transformers. In addition to these, brominated flame retardant is used in printed circuit boards, plastic castings, cables and cable insulation, which releases highly toxic dioxins and furans when burned to retrieve copper from wires. When the e-waste is used in land-fills, toxic substances leach out to contaminate the ground water; and when such waste is incinerated, toxic fumes are relased into the atmosphere which contribute to air pollution. There are, moreover, other serious environmental hazards of the electronic waste. The recycling of discarded computers, for example, has serious occupational and environmental implications. The glass panels and gasket in computer monitor, and the solder in printed circuit boards contain lead, which is known to damage the central and peripheral nervous system, blood, kidneys and the reproductive system in human beings even at low concentration. Lead adversely affects the endocrine system and hampers the development of brain in children. Similarly, cadmium compounds are also toxic to human beings and animals. Since cadmium is eliminated only very slowly from the kidneys, it accumulates in the kidneys; and once the critical concentration of cadmium is reached, there may be an irreversible kidney damage. It has been estimated that roughly 22% of the annual world consumption of mercury is used in electrical and electronic equipment. Mercury is highly toxic, and it can cause damage to various organs including brain and kidneys. Chromium is used as corrosion protection of untreated and galvanized steel plates, and also as a decorative or hardener for steel housings. Chromium easily passes through cell membranes and produces toxic effects. Chromium can also cause damage to DNA. Polyvinyl chloride (PVC) finds significant uses in electronics industries. Dioxin, which is known to be a carcinogin can be produced when PVC is burned. Barium is used in computers in the front panel of cathode ray tubes to protect the users of computers from harmful radiations. Short-term exposure to barium may cause swelling of the brain tissue, muscle weakness, and damage to heart, liver and spleen. Thus the list of toxic chemicals in the electronic waste seems to be almost endless.

A study conducted by *Toxic Links*, a Delhi-based NGO, has revealed the highly hazardous conditions under which the e-waste is recycled in India. Manual handling, burning in open air, acid baths and indiscriminate dumping of the e-waste are a common sight at any Indian waste-recycling facility. As a result, the workers and residents around these facilities are exposed to hazardous chemicals, and land, water and air are polluted. Due to high obsolescence rate of computers and other electronic items and the limited domestic recycling infrastructure in the US and other developed countries, dumping of used or obsolete electronic systems into developing countries has become the easiest option. It is evident, therefore, that strict national and international control and guidelines for the disposal of electronic waste are urgently needed.

8.3.6 Adverse Effects of Noise Pollution

8.3.6.1 Noise Pollution

When the phrase "environmental pollution" is mentioned, the thought it triggers in the mind of most people are usually those of air or water pollution. There is, however, another type of insidious pollution, known as "noise pollution", which adversely affects the physical and mental health of human beings and reduces their efficiency at work and enjoyment of life at home. Prolonged exposure to excessive noise may cause hearing loss, interference with speech, communication and sleep, reduction in working efficiency, irritation, fatigue, nervousness, nausea, increased blood pressure and many other physical and psychological symptoms. Noise pollution is no longer a monopoly of the highly industrialized nations of Europe and North America. As a result of the accelerated pace of industrialization and urbanization, the menace of noise pollution has raised its ugly head in many of the Third World countries too, and India is no exception.

Noise may be defined as the sound that is unpleasant and unwanted by the listener because of its bothersome nature, interference with the perception of wanted sound, or its harmful physiological and psychological effects. The presence of high levels of noise in the environment is regarded as a form of pollution because it lowers the quality of life. There are several ways in which excessive noise can affect the people adversely. The adverse effects of excessive noise can be classified into :

(1) Auditory effect,
(2) Physiological effect,
(3) Effect on task performance,
(4) Interference with sleep and
(5) Other effects.

8.3.6.2 Auditory Effect

Excessive noise can damage the human ear when a person is exposed to it over a long period. Even a short exposure to very high levels of noise may cause a temporary hearing loss. In fact, any continuous noise level higher than about 85 dB (decibels) can cause damage. This damage is slow, progressive and insidious; so it is not usually noticed by the victim (e.g., a worker in a noisy industry) till it is too late. The most prominent auditory effect of the high level of noise is the noise-induced hearing loss (or noise deafness). The sensitivity to noise varies from person to person. Some of the hearing loss resulting from the high-intensity noise is temporary and, over a period of time, partial or complete recovery will take place. Such a temporary loss of hearing is known as *"temporary threshold shift"*. On the other hand, any hearing loss that persists is known as *"permanent threshold shift"*. Once a permanent threshold shift has occurred, normal hearing cannot be restored. If a person is suddenly exposed to a very high level of noise, then the shift in the threshold of audibility may be permanent. This phenomenon is known as *"acoustic trauma"*.

8.3.6.3 Physiological Effect

In addition to the auditory effect, high level of noise has several harmful physiological and psychological effects. Some of the adverse physiological effects are:

(1) Change in blood pressure (usually an increase),
(2) Change in heart beat rate (usually an increase),
(3) Various glandular changes and
(4) Change in activity.

It has been found by various investigators that the frequency of cardio-vascular disease and hypertension (i.e., high blood pressure) is higher in industrial workers exposed to high levels of noise than in those who work in less noisy environments. Excessive noise, moreover, interferes with the proper functioning of the digestive system, especially when the noise is sudden and unexpected. Such a noise produces a decrease in the secretion of digestive juices, and also in the bowel activity.

8.3.6.4 Effect on Task Performance

Any new sound, or change in an existing and familiar sound, may result in atleast a momentary distraction of the worker. This distraction may impair the worker's ability to perform certain non-routine tasks. It has been observed by researchers in industrial engineering that workers are less accurate when working under noisy conditions. They are also likely to make mistakes which may lead to serious accidents. Industrial workers exposed to high level of noise are found to have higher incidence of circulatory problems, cardiac diseases, hypertension, peptic ulcers and neurosensory and motor impairment. All these serious diseases lead to increased absenteeism among workers and, therefore, reduce industrial productivity.

8.3.6.5 Interference with Sleep

The low-frequency sound can affect higher centers of the brain, which may cause disturbance in the normal sleep pattern and prevention of deep sleep. Older people are particularly affected by this disturbance of sleep. Once awakened by noise, they may not be able to go to sleep again that night. Noise has multiple effects on sleep. Aeroplane noise produces more disturbances in sleep than other types of noise. It is a common experience that noise reduces the depth and quality of sleep, which may adversely affect the overall mental and physical health. Noise disturbs sleep by causing more frequent awakenings. It has been found that an intermittent noise disturbs sleep more than a continuous noise.

8.3.6.6 Other Effects

High levels of noise also give rise to many adverse psychological effects. The mildest of such effects are physical and mental fatigue and lack of concentration. These mildest effects are quite important in industrial situations since they lead to lowered efficiency, increased absenteeism and higher rates of industrial accidents and injuries. Excessive noise is known to produce behavioural disturbances like annoyance, distraction, fatigue and speech interference. In a non-work environment, the psychological effects of excessive noise may lead to many psychosomatic disorders, tension-related diseases, mental illness and emotional distress. High levels of noise make human beings more irritable and more susceptible to irrational behaviour. This leads to higher rates of crime with the progress of industrialisation and urbanisation.

An important effect of high-level noise is its interference with speech and communication. Speech production and reception is the most important and also the most complex use of the auditory system in human beings. Noise can either mask the whole speech to make it inaudible, or it may mask only some of the frequencies present in it to make the speech audible but less intelligible. In either case, the process of verbal communication is hampered.

8.4 CONTROL OF ENVIRONMENTAL POLLUTION

8.4.1 Control of Air Pollution

8.4.1.1 Industrial Emissions

The extent of air pollution is very high in some of the Indian cities like New Delhi, Kolkata, Mumbai, Kanpur, Ahmedabad and Hyderabad, and in the steel cities (Bhilai, Bokaro, Durgapur, Jamshedpur, Rourkela and Visakhapatnam). Air pollution due to industrial emissions can be controlled by following some of the strategies listed below:

(1) Control at the Source: For the control of air pollutants at the source or their abatement, every industry (including thermal power plants) should segregate the collection of various particulates and the gaseous pollutants. Gas cleaning equipment should be used for the removal of particulate matters as well as for gaseous pollutants. The equipment used for the removal of particulates and aerosols is based on different principles, and its operation is based on the size of the particles . The purification of gaseous pollutants is achieved by using various methods of gas conversion like chemical conversion through oxidation, catalytic conversion, physical conversion (liquefying, cooling, and compressing the gas under pressure), and through various methods of gas removal such as adsorption on solid

media (activated carbon, clay, gel, silicates and aluminates), and absorption (dissolving different gases in different liquids).

(2) Process Modification: The industries, power plants, smelters, etc., should modify their plants, processes, technology and even raw materials in such a way that the production of air pollutants is as little as possible.

(3) Meteorological Control: For proper dispersion of air pollutants, industries should exercise meteorological control. This can best be achieved through proper design of stacks. The maximum concentration of any air pollutant at any point at the ground level is directly proportional to the mass rate of emission of that pollutant, and inversely proportional to the square of the height of the stack and the horizontal flow velocity of the air. It follows, therefore, that stacks should be of greater height and located in open areas where maximum wind velocity is available in order to minimize the ground-level concentration of air pollutants.

(4) Storing the Pollutants: A balancing storage should be created for storing the treated gases for a regulated release at uniform rate to achieve constant dilution of air pollutants with the atmospheric air. This strategy can avoid conditions of high pollutant concentration in the atmosphere at the time of peak discharge of the pollutants.

(5) Proper Location: Industries and power plants should be located at the outskirts of the cities, on higher hilly areas, and on the down-wind direction, in order to reduce the concentration of air pollutants in the city area.

(6) Stack Treatment: Conversion and removal devices should be installed for the stack treatment of gaseous pollutants. A good example of removal device is the electrostatic precipitator (ESP), which removes the particulate matters from stack emissions. An example of conversion device is the catalytic converter, which is installed in motor vehicles to reduce air pollution due to vehicular exhaust.

(7) Overall Control: Emphasis should be laid on an overall control of environmental pollution. The use of coal, oil and other highly polluting fuels should be minimised for domestic and industrial uses. Furnaces, kilns, smelters, etc., should not be permitted in the city area.

(8) Industrial Survey: A detailed industrial survey should be undertaken to identify the existing pollutant-generating industries situated within the city limits. As far as possible, highly polluting industries should be removed from the city areas and shifted to alternate sites on the outskirts of the cities in the down-wind direction.

(9) Regular Monitoring: The monitoring of air pollution should be carried out on a regular basis in order to facilitate timely preventive action on the

part of the industries and regulating authorities like Central and State Pollution Control Boards. There should be regular checking of the quality of ambient air and exhaust gases from motor vehicles at strategic locations of the city.

8.4.1.2 Vehicular Exhaust

Mass manufacture of motor vehicles has now spread in our country. This has greatly facilitated the mobility of people and the transport of goods, in addition to generating employment. As a result, there has been a tremendous increase in the number of motor vehicles in India in recent years. This, however, has not been an unmixed blessing. Increase in the number of motor vehicles has led to congestion, traffic jam, parking problems, air pollution due to emission of toxic and corrosive fumes, noise pollution and injuries and deaths due to accidents. The toxic exhausts from motor vehicles are among the major sources of air pollution, next only to thermal power plants. The ever-increasing vehicular traffic density poses a continued threat to the quality of ambient air. Some of the strategies to reduce air pollution due to vehicular traffic are:

(1) A phased programme for the coming ten to twenty years should be formulated, keeping in view the engine types, time frame and air quality standards, so that by a certain date, the vehicular exhaust is fully contained.

(2) There is no need to have two sets of standards for vehicular exhaust emission (one set for old vehicles and the other for new ones) since there is not much difference between the old and new vehicles as far as exhaust emission is concerned. This is the conclusion of a study conducted by the Indian Institute of Petroleum, Dehradun and a similar survey carried out by the Central Pollution Control Board. There should be only one exhaust emission standard for vehicles of all ages.

(3) The government fiscal policies should encourage updating the engine systems and installation of pollutants emission abatement equipment, matching the existing engine designs.

(4) The petroleum industry should seriously take up the supply of unleaded fuel and CNG (compressed natural gas).

(5) Existing roads and traffic systems should be improved so that different types of traffic (ranging from pedestrians, bicycles, rickshaws, animal-driven vehicles to two, three or four wheelers) can use the same road. This can lead to a smooth flow of traffic and reduce air pollution due to vehicular exhausts.

(6) Traffic laws and supervisory systems should be updated for better control of traffic in congested cities and towns.

(7) Government should encourage battery-operated public transport systems.

(8) Government should also encourage the use of CNG as a fuel for motor vehicles.

(9) Government and NGOs should promote the use of bicycles for short distances. The bicycle is a wonderful vehicle in many ways: (1) It does not pollute the environment, (2) It can carry more load than its own weight, (3) It runs on bio-energy, which is renewable and finally, (4) It provides the much-needed physical exercise to the rider.

(10) Government should encourage the use of public transport systems in preference to single-person vehicles.

8.4.1.3 *Thermal Power Plants*

Thermal power plants have the dubious distinction of occupying the highest position among the air-polluting industries. The pollutional aspects of thermal power stations have not yet been worked out in depth in our country. Thermal power stations cause environmental pollution right from the handling, loading, transport and unloading of vast quantities of coal to its burning for the production of steam required for power generation. Moreover, the quality of the Indian coal is inferior due to its high ash content. Hardly, 5% of the fly-ash generated by thermal power plants is utilized for making cement, bricks, etc., and the rest of it is dumped in ash disposal areas, which itself creates many environmental problems. Although Indian coal is fairly low in sulphur content (about 0.5%), it has been estimated that our thermal power plants release about one million tonne of sulphur dioxide (SO_2) per year, which contributes to air pollution and acid rain. In addition to fly-ash and SO_2, thermal power plants release vast quantities of particulates, hydrocarbons, carbon monoxide (CO), various oxides of nitrogen (NO_x) and various hazardous trace elements. In order to reduce and control the environmental pollution due to thermal power plants, the following steps have to be taken:

(1) Pollution control technology should be utilized on a larger scale. It should either be developed domestically or imported and adapted. At present, very few thermal power plants have installed the requisite ESPs for the removal of particulate matters. Even when installed, the ESPs do not function properly in many cases due to the power plants being old in design and using coal with high and variable ash contents. We urgently need low-cost technologies for the removal of SO_x, NO_x, CO, hydrocarbons and other air pollutants emitted by thermal power plants.

(2) State Pollution Control Boards and other agencies should carry out the Environmental Impact Assessment (EIA) for all the existing as

well as proposed thermal power plants. This applies particularly to the Super Thermal Power Plants like the power plants at Singrauli and Talcher. The EIA should be particularly related to proper siting, stack height, and maintenance of operational and control equipment.

(3) There is an urgent need to generate reliable analytical data on coal-based thermal power plants on a national basis.

(4) The country should establish major eco-friendly facilities for washing and beneficiation of coal.

(5) The Ministry of Energy, Government of India, should establish a major R & D (Research and Development) facility to develop clean technologies related to power generation, and to explore non-conventional sources of energy.

(6) Our research laboratories and technological universities should be encouraged to carry out R & D work to develop technologies for utilization of fly-ash on a much larger scale.

8.4.1.4 *Some Other Strategies*

(1) For controlling any type of pollution (like air, water or noise pollution) at its source, the most important prerequisite is to establish industry with specific Minimum National Standards (MINAS). There are about 20 broad types of industries in India, and such documents have already been prepared for more than half of these groups of industries. For quick implementation of MINAS, the central as well as the State Pollution Control Boards have to be strengthened with an adequate number of properly trained manpower. At the same time, the industries should be given appropriate incentive for adopting pollution control measures.

(2) Our cities are the major sources of environmental pollution; so conscious decisions have to be taken to disperse industries rather than aggregate them in class I and class II cities. At present, almost 75% of our industrial production is generated from nine metropolitan cities.

(3) The selection of a particular manufacturing process is of great importance, and due consideration should be given to that. If the production technology is selected with great care, it would make tremendous difference to both the end product and the pollutional costs to the society. For example, caustic soda can be produced today by using mercury cell, diaphragm cell or membrane cell. Among these three technologies, the use of membrane cell eliminates the use of mercury (a very toxic pollutant), and also reduces the cost of pollution control.

(4) India is a country where centuries co-exist. Thus we have situations where a pulp and paper mill with a hundred-year old, highly-

polluting batch digester technology is going on, side by side with a modern continuous-feed digester technology. Similarly, hundred-year old looms are in operation today along with the most modern ones. In India today, there is no premium on pollution-free modern technology. Conscious decisions have to taken, therefore, to do away with old and polluting technologies in preference to modern and environmentally clean technologies.

(5) There is an urgent need to review and reinforce the criteria for industrial site selection, procedures for *"No Objection Certificate"*, and regular reports on the progress made with the installation of pollution control measures of new industries. Only then, can we ensure that we do not add to our already huge backlog of pollutional load.

(6) Chronic power shortage in our country today is coming in the way of proper pollution control. On the pretext of power shortage, industries shut off their treatment plants in preference to production units. Obviously, there is an urgent need for developing domestically or importing low-energy technologies for pollution control.

(7) Earlier, it was a custom (particularly in small towns) to locate the commercial activity on the ground floor and to live on the first floor of the house. The purpose was to live very close to one's work place (a shop or a tiny industrial unit). Unfortunately, this old tradition is still being followed in many of our towns and cities, often with highly polluting industries like electroplating and recycling of plastics. Steps should be taken by authorities to segregate the work place and living quarters.

(8) Finally, India should intensify its R & D activities in the whole area of pollution control. This involves many disciplines including various branches of engineering (like chemical, civil and mining engineering), basis sciences (like chemistry, physics and biology) and many inter-disciplinary areas like biotechnology. Side by side, we should also strengthen our database on pollution control and all the related aspects.

8.4.2 Control of Water Pollution

8.4.2.1 Self-Purification

When oxidisable matters such as sewage or sullage are discharged into a river or other water bodies, they get decomposed by micro-organisms. This process creates biochemical oxygen demand (BOD) in water and therefore, reduces its dissolved oxygen (DO), which is an important parameter to judge the quality of water. On the other hand, the organic matter which is biochemically non-oxidisable creates chemical oxygen demand (COD). Normally, in a healthy water body, there is enough dissolved oxygen to support biochemical oxidation of the biodegradable organic matter. This process is known as self-purification. In the absence of an adequate amount of dissolved oxygen, there is anaerobic decomposition of organic matter, which gives rise to septic conditions and foul smell. In fact, the relationship between DO and BOD is one of the important leads to assess the self-purification ability of a river system or other water bodies. In case of the river Ganga, for example, the Central Pollution Control Board came to the conclusion (based on the data available on the DO and BOD of the river) that for pollution at a single point like Kanpur, it should not take more than a day, or about 100 km downstream, for the pollutants to be removed by the process of self-purification of the river. This, however, does not happen in practice since rivers usually receive, apart from point pollutional loads, a significant quantum of distributed pollutional load brought down by tributaries, local contributions, runoffs, etc. As a result, self-purification may be regarded only as a "First Aid Strategy" for the control of water pollution.

8.4.2.2 *Treatment and Disposal of Sewage*

In developing countries, the organic pollutional load (contributed mainly by municipal sewage) is a major problem in connection with water pollution. In India, it has been estimated that municipal sewage contributes 90% organic pollutants to our river systems, 7% is contributed by large and medium industries, while the remaining 3% comes from small-scale industries. Thus treatment, utilization and proper disposal of municipal sewage are very important steps towards the control of water pollution. Some of the important methods used for the disposal of sewage are as follows :

(1) Land Disposal and Sewage Irrigation. In this method, the sewage is disposed off on land (often for the purpose of irrigation) instead of discharging it into a water body. In this case, the land area acts as a crude filter and stabilizes the sewage by aerobic filtration. The sewage is usually given primary treatment before its disposal on land. This method is very similar to intermittent sand filtration. In case of sewage irrigation, the water and the fertilizing elements in the sewage, viz., nitrogen, phosphorous and potassium, are utilized to raise crops. Hence, this method has the special advantage of fertilizing the land.

(2) Dilution. In this method, the sewage is discharged into a large body of water like river lake or sea. The sewage is usually treated before discharging it into a water body to ensure that the condition of the receiving water body does not deteriorate to the extent of impacting on its normal use. The cities and towns which are situated near a large river (Allahabad, Kanpur, Varanasi and Patna, all of which are situated near the river Ganga) mainly use the method of dilution to dispose off their sewage. This method is considered quite satisfactory during the rainy season when the dilution

factor is extremely high, but not suitable for the period from November to June, when the volume of river flow or the volume of water in the river is low.

Treatment of Sewage

Primary Treatment: The primary treatment of sewage mainly consists of physical separation of floating and suspended solids (that are settleable) from the sewage. The main equipment used for the primary treatment are screens, grit chambers, detritus tanks, skimming tanks and settling tanks.

Secondary Treatment: The secondary treatment of sewage includes flocculation and precipitation of the remaining materials in the sewage with the help of biological agencies, and their physical separation in secondary settling tanks. The by-products of the secondary treatment process are screenings, grit and sludges of different kinds. Out of these, the first two can be easily disposed off by burial or burning. Sludge, on the other hand, contains unstable volatile organic substances; so it may be treated by the method of digestion. In the process of digestion of sludges, various gases are produced with high calorific value which can be successfully utilized for heating or power generation. The solid materials left behind (i.e., the digested sludge) contain many fertilizing materials and are useful as manure or soil builder.

Sludge Digestion: The process of sludge digestion consists of liquefaction of organic materials in the sludge by anaerobic bacteria, which produce an alkaline reaction. When the sludge is first placed in the digesting tank, acid digestion with the production of noxious gases results and eventually, alkaline digestion prevails. Once established, the alkaline condition remains in the digestor for an indefinite period. The gases given off by the sludge are mainly methane (CH_4) with some carbon dioxide (CO_2) and small quantities of other gases. Among the sludge gases, methane has a calorific value of 8,000 kcal/m^3. Since methane forms about 67% of the sludge gases, the calorific value of the sludge gas can be taken to be approximately 5,334 kcal/m^3. The sludge gas occupies about 0.95 m^3/kg of volume at normal temperature and atmospheric pressure. The main use of the sludge gas, apart from heating the digestion tanks, is for generating power. The power thus generated is enough to pump the entire sewage. Dried digested sludge can be used as a fertilizer since it contains 0.8% to 3.5.% nitrogen, about 1.6% phosphorous and 0.35% potassium on the basis of the dry weight of solids. The digested sludge may also be used for land filling, incinerated to produce heat and power, or dumped into large bodies of water.

8.4.2.3 Treatment and Disposal of Industrial Effluents

As a consequence of rapid industrialization in India in the post-independence period, the problem of environmental pollution due to industrial effluents has grown significantly. Since a large majority of industries use vast quanties of water, considerable amount of waste water emanates from them. The problem gets more aggravated since the industrial effluents are as varied in their characteristics as the industries themselves. No standard procedure applicable to all types of industrial effluents can be recommended. For a variety of reasons, industrial effluents are usually discharged into water bodies, either inadequately treated or never treated. This creates the problem of surface and subsoil water pollution. Thus the pressure on water resources due to rapid industrialization is two-fold :

(1) Requirement of large quantities of water by industries and
(2) Non-availability of clean water because of its pollution due to the discarge of industrial effluents.

It becomes imperative, therefore, that industrial effluents are treated adequately before they are discharged into water bodies.

Treatment of Effluents: Treatment of industrial effluents has to be handled in a systematic and planned manner. The steps necessary to achieve this purpose are :

(1) Analysis of each effluent with its rate of flow (to determine the daily, monthly and yearly average volume of the effluent concerned),
(2) Identification of various pollutants and determination of the total pollutional load (with the help of the flow rate) in different effluents,
(3) Separation of concentrated effluents, and separate planning for the treatment of concentrated and dilute effluents and
(4) Use of material balance with regard to raw materials, intermediates and finished products to verify the amounts of various pollutants in different effluents.

The methods used for the treatment of industrial effluents can be classified into three main categories: (1) physical treatment, (2) chemical treatment and (3) biological treatment. These methods are briefly described below :

Physical Treatment: The physical treatment of industrial effluents employs various devices:

(1) Coarse or fine screens as well as bar screens (to remove materials such as pieces of bricks or wood, rags, paper, cloth, etc.);
(2) Devices like grinders, cutters or shredders (to break up solid material),
(3) Grit chambers (to remove sand, dust, cinders and other inorganic settleable materials),
(4) Grease traps (to remove unemulsified oil and grease) and
(5) Sedimentation tanks (to primarily remove suspended organic solids from the effluents prior to their biological treatment).

Chemical Treatment: The chemical treatment of industrial effluents involves processes like neutralization, coagulation and flocculation and chemical destruction of toxic substances like cyanides. The chemical treatment is aimed at removing the dissolved materials from the effluents. The most important solutes are the compounds of phosphorous and nitrogen. Among these, phosphorous is more important than nitrogen in causing eutrophication of water bodies, which is a process resulting in excessive and undersirable growth of plant life in lakes and harbours. Other toxic substances may also be present in industrial effluents and special processes have to be used for their removal. Phosphorous can be removed by precipitation with coagulants like $Al_2(SO_4)_3$ and F_eCl_3. Lime, $Ca(OH)_2$, can also be used to precipitate phosphorous. Nitrogen, on the other hand is more difficult to be removed than phosphorous; but when it is in the form of ammonia (NH_3), it can be stripped by air at pH=11.0. Apart from phosphorous and nitrogen, there are many other dissolved substances present in industrial effluents. Some of them can be removed by filtration through certain media (such as activated charcoal, which can be reactivated and reused). Many of these substances can theoretically be removed by the biological treatment. Care must be taken, however, not to disturb the chemistry of the effluent under treatment to such an extent that the beneficial micro-organisms are harmed.

Biological Treatment: As the name indicates, the biological treatment harnesses natural fauna and flora for the treatment of waste water. Since most of the industrial effluents are deficient in one or the other of essential nutrients (nitrogen, phosphorous and potassium), it is always advantageous to biologically treat industrial effluents in conjunction with municipal sewage, which provides not only these nutrients, but also a heterogeneous population of bacterial flora required for the breakdown of various pollutants.

A variety of operations, either singly or in combination, may be employed for the biological treatment of industrial effluents. The choice of the method of treatment is determined by various factors :

(1) The characteristics of the effluents to be treated,
(2) Circumstances pertaining to the body of water being used for the final disposal of effluents,
(3) Requirements of the State Pollution Control Board (or other regulatory agency) and
(4) The cost of the treatment plant and its operation.

The anaerobic treatment process has been found to be very effective for effluents containing high concentrations of organic matters (e.g., effluents from distilleries, tanneries and slaughter houses). The aerobic process, on the other hand, is particularly useful for industrial effluents which are not very rich in organic matter, e.g., dairy effluents. Quite often, the anaerobic treatment process has to be followed by the aerobic process for the complete treatment of waste water.

Disposal of Industrial Effluents: The treated industrial effluents can be discharged into surface water bodies (like streams, rivers and lakes), sewers, or on land. The mode of disposal has to be selected with great care, on the merit of each case, to ensure that the problem of environmental pollution, if not totally eliminated, is minimized. Moreover, some of the treated industrial effluents have a good potential as fertilizers, and these should be exploited whenever possible to recover some of the treatment costs. This practice will also be beneficial to the soil; but it has to be ensured that the treated industrial effluents are safe enough for irrigation, and they do not contain toxic metals like Hg, Pb and Cd beyond the recommended limits.

8.4.3 Control of Soil Pollution

8.4.3.1 *Control of Solid Waste*

Generally speaking, solid wastes are classified into two broad categories, viz.,

 (1) Municipal solid waste and
 (2) Industrial solid waste.

   Municipal (or household) solid waste is generally collected locally, and its quantity depends on the size and consumption habits of the local population. It is disposed off on either public property or, in some cases, on private contractors' sites. Industrial wastes are those generated by industrial firms. The detailed statistics for the generation and disposal of solid wastes are generally not available for developing countries. A scrutiny of the composition of solid wastes, both in developing and developed countries, clearly shows that the percentage of paper, rubber, leather, plastics, metals and glass increases considerably with increasing industrialization of a country due to the wide use of these materials in daily life.

   There are three methods commonly used for the disposal of solid wastes:

 (1) Sanitary Land-fill,
 (2) Incineration and
 (3) Composting.

Sanitary Land-Fill: In developing countries, the sanitary land-fill technique is often confused with the *"open dump"*, but dumping of garbage in the open is not a sanitary land-fill. There are five principles which should be strictly adhered to in the case of a sanitary land-fill. These principles are :

 (1) Solid wastes must be deposited in a controlled manner.
 (2) Solid wastes should be spread in thin layers.
 (3) A ground cover of not less than 15 cm should be applied daily.
 (4) There should be no open burning of solid wastes.
 (5) All the factors which are likely to allow the deposited wastes to contribute to water pollution must be eliminated.

Planned management with respect to sanitary land-fills begins with the analysis of soil conditions surrounding the proposed site. A ravine or a man-made gravel pit is generally sought as a sanitary land-fill site. Soil conditions surrounding the land-fill site are quite significant in the ultimate success of the land-fill. Fine-grained soils are more useful than coarse-grained sandy soils in containing the movements of unwanted gases and water outside the land-fill area. The type of soil also ultimately determines how the land-fill will be utilized, once it is completely filled.

Incineration: Incineration of solid wastes serves the function of reducing the volume of these wastes from their raw states to more manageable levels, thus reducing the transportation costs to the ultimate disposal site, and accommodating the wastes of a larger population for a given area available for the land-fill. The two main disadvantages of this method are that both capital and operating costs of incinerators are high, and incineration of wastes contributes to air pollution. Two types of incinerators are generally used for unsorted wastes. These are known as the *batch-type plant* and the *continuous-feed plant.* The batch type plant is manually stoked and has a relatively small-rated capacity. Since the operation of the batch-type plant is intermittent, there is a lack of uniform burning temperature. This promotes inadequate combustion and, therefore, a heavy output of particulate matter, less than optimal volume reduction, and an unstable residue. The batch-type units are generally unsuitable for large urban centers, although they are used in the smaller communities. The continuous-feed plant has larger storage bins, automatic feed hoppers and a variety of ash removal systems. The unit maintains uniform combustion temperature range, can be fitted with pollution control devices, and yields a stable residue. The reason the technique of incineration is experiencing rapid growth is the large volume reduction that can be achieved in a well-designed incinerator with controlled furnace temperature in the range of 1,400-1,800 °F. With increasing population, land is growing increasingly more scarce, more expensive and available only at a distant place from waste collection points. Thus the importance of incineration method for solid waste disposal lies in the fact that incineration cuts the land requirement to about one-third of that required if the wastes were to be land-filled directly, i.e., before being incinerated first. Moreover, a stable residue obtained after incineration cuts drastically, or eliminates entirely, the need for covering daily the land-fill with a cover material.

Composting: (Re: 9.1.4.4.)

Gobar Gas Plants: In India, about 30% of the energy consumed by people is biological in nature. The *gobar gas plants* (a popular name for bio-gas plants using cattle dung to produce bio-gas and manure) are based on anaerobic fermentation of organic wastes. These plants increase the heating efficiency of the cattle dung by about 20% and produce an organic manure which is about 43% better than dry cattle dung itself. This manure, in addition to fertilizing the soil, can also reduce the pollution of soil and water caused by naphtha-based fertilizers.

The *gobar gas* is a mixture of methane (about 60%), carbon dioxide (almost 35%) and a very minute quantity of a few other gases. As a fuel, the gobar gas is cleaner (from the point of view of air pollution) than the dry cattle dung itself. Bio-gas is a good answer to the deepening problem of energy crises in India especially the energy needed for cooking, lighting and heating in rural areas. The reason is that bio-gas is not only a non-polluting source of energy; it also provides enriched manure and improves sanitation in rural India. In our country, the design of gobar gas plants of various capacities has been standardized by the Khadi and Village Industries Commission (KVIC) and other agencies. As a result, thousands of gobar gas plants have been installed all over the country to meet the increasing demand of energy in rural areas.

8.4.3.2   *Control of Liquid Wastes*

In addition to domestic and industrial solid wastes, soil is also polluted by liquid wastes like municipal sewage and industrial effluents. The municipal sewage can be decomposed readily by natural processes, i.e., by the action of micro-organisms (like bacteria, fungi, protozoa and other microbes). Biodegradable municipal sewage can also be treated by artificial methods in a sewage treatment plant. Serious problems arise, however, when non-biodegradable pollutants enter into the soil through liquid wastes like industrial effluents. Moreover, there is a limit to the total amount of waste that can be decomposed in a given area of land by natural processes.

An important method used to process the domestic sewage is the *septic tank*, which is an underground sewage container made of concrete. Waste materials settle at the bottom of the tank and form a sludge, while the liquid flows out of the tank into a system of perforated pipes. The stream of liquid wastes passes slowly through the holes in these pipes and gets filtered by the soil. The sludge, which accumulates at the bottom of the septic tank, is slowly digested by the micro-organisms in the sewage. The sludge is periodically pumped out and transported to a sewage treatment plant for the final treatment and disposal. Handling and disposal of sludge is a major problem faced by sewage treatment plants. The main worry in this connection is: where to dump the ever-increasing amount of digested sludge? Due to acute shortage of suitable dump sites, it has been suggested that the sludge, after undergoing microbial decomposition of its organic matters, should be pumped into strip-mined, barren, or marginally productive land, away from the town or city where the sewage is generated. This method requires only about one-third of the cost of sewage treatment by the conventional method and helps recover strip-mined or barren land for productive uses.

8.4.3.3 Recycling of Complex Soil Pollutants

Complex soil pollutant include paper, glass, plastics, metals and non-biodegradable organic matters. Recycling consists of recovering useful materials from the waste and reusing them.

Paper: (Re: 9.3.3.2.)

Glass: (Re: 9.3.3.3.)

Plastics: (Re: 9.3.3.5.)

Metals: (Re: 10.5.2.)

Organic Matter: Organic materials contained in solid wastes can be subjected to aerobic decomposition by micro-organisms. This process leads to formation of compost which acts as an excellent conditioner of soil. In developed countries, however, inorganic synthetic fertilizers are cheaper and more convenient; so they are preferred over the compost. In India and in many other developing countries, the use of compost is very common.

The giant steps taken recently by biotechnological research (especially in the area of genetic engineering) have the potential to help reduce the pollution of soil. At present, several genetically engineered strains of bacteria are available to decompose the complex organic compounds that were considered non-biodegradable previously.

8.4.4 CONTROL OF NOISE POLLUTION

8.4.4.1 Noise Control Methods

Noise control is the technology for obtaining an acceptable level of noise at one or multiple receivers, taking into account the economic and operational aspects of the situation. Since each industrial noise control problem is usually different from others, it should be treated on an individual basis. Various techniques adopted for noise control can be classified into three broad categories:

(1) Control at the source.
(2) Control of the transmission path.
(3) Control at the receiver.

Control at the source is the best method to control noise pollution. In industrial situations, noise control at the source can be realized by using one or more of the following three techniques:

(1) Reduction of the amplitude of vibrating parts (since the intensity of the resulting noise is proportional to the square of the amplitude).

(2) Reduction of the response of various components to the exciting forces.

(3) Changes in operating procedures (to reduce the number of vibrating parts).

Control of the transmission path involves the following techniques :

(1) Proper location of the industry.
(2) Proper layout of the building.
(3) Path deflection of noise.
(4) Providing enclosures.
(5) Absorption of sound (along the path to the receiver).
(6) Impedance mismatch to avoid resonance.

Control at the receiver is achieved by using one or more of the following methods:

(1) Use of personal protective devices, e.g., ear plugs.
(2) Exposure control (i.e., limiting the period of exposure).
(3) Education and publicity (i.e., creating awareness on the hazards of excessive noise).
(4) Job rotation and provision of quieter working areas for sometime.
(5) Regular monitoring of the hearing levels of personnel working in noisy areas.

Community Noise: From the point of view of noise pollution, the community is divided into the following four categories :

(1) Industrial areas.
(2) Commercial areas.
(3) Residential areas.
(4) Silence zones (such as schools and hospitals).

Some methods for controlling noise pollution in the above areas are indicated below:

Industrial Areas:

(1) Noise generated by various machines should be controlled by proper designing and maintenance.
(2) In case of noisy machines, noise barriers should be provided to protect the workers from excessive noise.
(3) Operators of noisy machines or those working in noisy areas of the factory should be provided enclosures.

Commercial Areas:

(1) The volume of traffic should be reduced by diversion of it.
(2) Parking of vehicles should be restricted to a few areas only.

(3) Use of noise generating systems (e.g., music systems in restaurants and shops) should be either prohibited or their volume should be kept low.

(4) Where generators are used, they should be provided with acoustic hoods to absorb some of the noise produced by them.

(5) Unnecessary honking of horns by vehicles should be prohibited in commercial areas.

Residential Areas:

(1) Movement of heavy vehicles like trucks and buses should be restricted in residential areas, and should be totally prohibited during the night.

(2) Use of horns by vehicles passing through residential areas should be restricted.

(3) Residential areas should have plenty of trees and bushes in open spaces, houses and lanes to absorb sound.

Silence Zones: Areas upto a distance of 100m around certain premises (like hospitals, nursing homes, educational institutions and courts) may be declared as silence zones by competent authorities. Some of the noise control measures in the silence zones are:

(1) Plying of heavy vehicles like trucks and buses should be banned in and around the silence zone areas.

(2) Use of horns should be banned inside silence zones.

(3) There should be uninterrupted supply of power in silence zone areas, so that the use of noisy generators can be avoided.

8.4.4.2 Creating Public Awareness

The best method to control noise pollution is to *control at the source.* In our country, noise pollution in urban areas is generally caused due to indiscriminate use of loudspeakers, fire-crackers, frequent use of loud horns and poor maintenance of vehicles. Industrial noise pollution is mostly confined to industrial premises; but when industries are wrongly located and improperly planned, they may create the problem of noise pollution and affect the people residing in the neighbourhood.

In contrast to the industrial noise, which affects mostly the workers of the industry concerned, noise produced by the commercial sectors in our country largely affects the general public. People use loudspeakers indiscriminately in various social and religious functions for advertisements and on various occasions like birthdays and other anniversaries. Fire-crackers are used without any restriction to celebrate various religious functions. Due to lack of proper planning, the commercial areas in many cities and towns have grown within residential areas. This is the main reason why the general public is so much affected by the noise produced in the commercial sector.

As a matter of fact, the problem of noise pollution is increasing day-by-day due to lack of public awareness and education. We may not be able to achieve the best results in the area of noise pollution control, even with the best and strictest legislation in the world, unless we receive full co-operation from the public. Receiving such a co-operation is not possible unless the public is fully aware of the hazards of noise pollution. In order to generate public awareness, sincere efforts must be made by the government agencies, voluntary organization, educational institutions and the print and electronic media.

8.4.4.3 *Action by the Town Planning Authority*

(1) The residential, commercial and industrial areas and the silence zones in a city or town should be planned in such a manner so that the noise created in different zones is confined to the zone concerned and does not affect other zones. Any unplanned growth or encroachment should be removed at once.

(2) Establishment of industry or development of commercial activity should not be allowed within residential areas. The local bodies should deal with this problem firmly.

(3) Preferably, houses should not be located near the main roads. Attention should be given to the architectural layout of residential localities so as to reduce the transmission of noise from one house to another and also to reduce or eliminate reverberations in a residential colony.

(4) There should be plenty of trees and bushes in the open spaces between houses and lanes.

(5) The pavements should have minimum covering by stone and concrete slabs. As far as possible, use should be made of grass cover, instead of stone slabs or concrete, to facilitate absorption of noise and to avoid its reflection.

(6) Diesel generator sets used in buildings should have proper noise control system. They should be installed in closed rooms.

(7) In consultation with the Railway Authority, the Town Planning Authority should make the arrangements for erecting acoustic barriers near railway lines, reducing the speed of trains and avoiding whistling within the municipal limits, as far as possible.

(8) Acoustic barriers should be placed near construction sites during the on-going construction activities.

(9) There should be fencing around construction sites to prevent people coming near the sites and being exposed to unnecessary noise.

(10) Temporary earthen walls should be constructed near the construction site using the soil which is normally hauled away from the site.

(11) Acoustic walls should surround the buildings of hospitals, libraries, educational institutions and courts, which are situated near the main road.

8.4.4.4 Action by Airport Authority

(1) Aerodromes should be located far away from the city and growth of the city should not be allowed to extend upto the aerodrome.

(2) Aeroplanes should take off in the direction radially away from the city, which can be ensured by proper planning of runways.

(3) During boarding and unboarding operations, the planes should be parked sufficiently away from the airport building.

(4) Night-time landing and take-off operations of aircrafts should be minimized.

(5) During maintenance and repair of aeroplanes, the workers should use hearing protection devices like ear plugs.

(6) Portable silencers should be used in the plane at the intake as well as at the exhaust during the idling period at the airport.

8.4.4.5 Action by Town Administration Authority

(1) Town Administration should ensure that all parties intending to use loud-speakers or public address systems on any occasion obtain licence and follow the recommended guidelines while using them.

(2) Public address systems and loud-speakers should not be used at night (between 9.00 pm and 6.00 am), except in closed premises.

(3) Loud-speakers should be directed at the audience and not towards the neighbourhood.

(4) Loud-speakers should not be allowed for advertisement, other commercial activities, election campaigns and religious functions, atleast not during the night.

(5) The permitted strength of power amplifiers should be just adequate and the noise level beyond the boundary of the noise-source premises should not be increased by more than 5 dB (decibels) above the ambient noise level.

(6) Manufacture and sale of fire-crackers producing an impulsive noise of more than 90 dB at 5m from the site of bursting should be banned.

(7) Manufacture and sale of joined fire-crackers should be banned.

(8) Bursting of fire-crackers during the night (between 9.00 pm and 6.00 am) should be banned.

(9) Bursting of fire-crackers should be permitted only during public festivals.

(10) In commercial areas, the volume of road traffic should be reduced by its diversion; parking of vehicles in commercial areas should be restricted to a limited number of parking lots only.

(11) In commercial areas, the noise generating sources (like the music systems in various shops) should be preferably kept at low volume. Advertisement of articles for sale through loud-speakers should be avoided.

(12) Use of pneumatic (or pressure)horns in motor vehicles should be banned.

(13) Industrial workers working inside the industry should not be exposed to noise levels higher than that prescribed by the regulatory authorities (usually 80 or 85 dB).

(14) Sirens should not be allowed on any vehicles except for fire-brigade vehicles, ambulances and a few other vehicles used for emergency works. VIPs (very important persons) should not be permitted to use sirens on their vehicles.

8.4.4.6   *Action by Transport Authority*

(1) The movement of vehicles along the inner arterial roads should be restricted in residential areas.

(2) The vehicles should not be allowed to generate noise levels higher than the upper limit prescribed by the regulatory authorities.

(3) There should be restriction on the use of horns by vehicles passing by the residential areas and silence zones.

8.4.4.7   *Action by Pollution Control Boards*

(1) Standards of noise levels prescribed for various situations should be reviewed and revised from time to time.

(2) Massive awareness and education programmes targeting the general public and various NGOs (non-government organizations) should be organized frequently.

(3) Regular monitoring of noise levels should be carried out and status reports should be prepared.

8.4.5   PROTECTION OF OZONE LAYER

8.4.5.1   *The Ozone Layer*

After the greenhouse effect and the resulting global warning, the second global environmental problem concerns the destruction of the stratospheric ozone layer. The ozone (O_3) is a variant of oxygen (O_2), which contains three oxygen atoms instead of the usual two, as in the case of the oxygen molecule. Like oxygen, ozone too is a gas at ordinary temperature. At the ground level, ozone is an air pollutant since it strongly irritates the eyes and the respiratory system. Ozone, moreover, is a major component of the photochemical smog (a combination of smoke and fog, observed in highly polluted cities), and it may also act as a greenhouse gas.

In stratosphere (a layer of the atmosphere lying at a height of 10-20 km above the surface of the earth), there is a layer of low-density air, which contains approximately 300-600 ppb (parts per billion) of ozone. This ozone prevents the ultraviolet component of the sunlight (i.e., the solar radiation with wavelength less than 280 nm) from reaching the earth's surface. In fact, ozone is the only component of the atmosphere that absorbs the solar ultraviolet radiation significantly at the wavelengths below 280 nm (1 nm = 10^{-9} m). If the ozone layer in the stratosphere were removed, we would expect large amount of the solar ultraviolet radiation in the range 200-280 nm to reach the surface of the earth.

The energy of a photon of light is inversely proportional to its wavelength. As a result, the photons with shorter wavelengths (e.g., those of the ultraviolet radiation) are more chemically active than the photons of longer wave lengths (e.g., those of the visible or infrared radiation). If the high energy photons of the solar ultraviolet radiation reach the surface of the earth, they will cause chemical reactions in the surfaces they strike (including the human skin). These high-energy photons are, therefore, expected to cause increased rate of skin cancer in human beings. People may be able to protect themselves from the solar ultraviolet radiation by using sunhats and sunscreens, but plants and animals cannot. Sunscreens contain chemicals that are largely opaque to the ultraviolet light. It is not known how plants and animals would respond to the solar ultraviolet radiation if it reaches the surface of the earth. Thus the ozone, which is a harmful air pollutant at the ground level, acts as a beneficial shield for the harmful ultraviolet radiation in the stratosphere.

8.4.5.2 *Destruction of the Ozone Layer*

In the stratosphere, the destruction (or depletion) of the ozone layer is mostly caused by elemental chlorine atoms. The mechanism for this destruction of ozone involves the following two reactions :

$$O_3 + Cl \longrightarrow ClO + O_2,$$
$$ClO + O_3 \longrightarrow Cl + 2O_2.$$

Of course, there are some other chemical reactions going on in the stratosphere that modify and compete with the above two reactions. But if we ignore the other reactions and add the above two reactions, then we find that the net result is the following reaction :

$$2O_3 + Cl \longrightarrow 3O_2 + Cl.$$

Thus, in the presence of chlorine atoms (Cl), two ozone molecules are converted into three oxygen molecules, while there is no net consumption of Cl atoms, which merely act as a catalyst. As a result, one Cl atom can go on converting many ozone molecules (O_3) to ordinary oxygen molecules (O_2) before it is consumed by some other chemical reactions. It has been estimated that a single chlorine atom in the stratosphere can convert as many as 10^4 to 10^6 O_3 molecules to O_2 molecules before being consumed by some other reactions. This mechanism is often referred to as the catalytic destruction of ozone since the chlorine atoms act as a catalyst for the reaction.

Most of the chlorine in the world is in the form of chemically stable sodium chloride (NaCl), either dissolved in the oceans or in underground salt deposits formed by the evaporation of ancient oceans. Elemental chlorine (Cl), which is a very reactive chemical, has a very short life-time in the lower atmosphere and has few natural ways to get from the lower atmosphere up to the ozone layer in the stratosphere. The only naturally occurring chemical that can transport much chlorine high enough into the stratosphere to damage the ozone layer is methyl chloride (CH_3Cl), which is produced in large quantities by biological processes in shallow oceans. Most of this methyl chloride is destroyed in the troposphere (the lower atmosphere), but approximately 3% of the world-wide methyl chloride emissions reach the stratospheric ozone layer. Chemically active solar ultraviolet light in the range of 200 to 280 nm, which enters the ozone layer but does not penetrate below it, is strong enough to split up the methyl chloride molecules, releasing Cl atoms, which initiate the reactions given earlier and destroy the ozone molecules. Before we had the synthetic halogen compounds known as chlorofluorocarbons (CFCs), methyl was probably the main natural destroyer of the ozone layer. However, this destruction of the ozone was in balance with its natural production mechanisms, leading to a steady-state ozone layer.

Starting about a hundred years ago (i.e., from the first decade of the twentieth century), humans began releasing into the atmosphere synthetic chlorine-containing compounds in significant amount. The compounds like methyl chloride (CH_3Cl) that also contain hydrogen atoms can be attacked in the atmosphere by the OH radical. As a result, most of such chlorine-containing compounds do not survive long enough in the atmosphere to reach the stratosphere. Carbon tetrachloride (CCl_4), on the other hand, has no hydrogen atom; so most of it is believed to reach the stratosphere and participate in the destruction of the ozone layer. The world production of carbon tetrachloride, however, has remained more or less constant over the last 70-80 years.

8.4.5.3 Chlorofluorocarbons

The chlorofluorocarbons (CFCs) are a group of synthetic chemicals obtained from hydrocarbons by replacing hydrogen atoms in the latter with chlorine

(Cl) and fluorine (F) atoms. The CFCs were developed for the first time in 1930 in the USA by Dr. Thomas Midgely, a research scientist working at General Motors, for applications in refrigeration. Before development of CFCs, the most popular refrigerants were ammonia (NH_3) and sulphur dioxide (SO_2), both of which were toxic and corrosive. In contrast, the CFCs (marketed under the brand name "Freons" by the chemical giant Du Pont) were non-toxic and inert. When he synthesized the first CFCs in 1930, Dr. Midgely was hailed as a hero. He flamboyantly demonstrated to the American Chemical Society his synthesis of a non-toxic and non-flammable substitute for the refrigerants ammonia and sulphur dioxide by inhaling a lungful of CFC-12 (a trade name for CF_2Cl_2) and then blowing out a candle. The so-called "safe" synthetic chemicals created a boom in the refrigeration and air-conditioning industries in the decades following 1930s; but today the CFCs are among the most hated chemicals all over the world.

This shift in attitude towards the CFCs started in 1974 when two American scientists, Dr. F.S. Rowland and Dr. M.J. Molina in California, discovered that the stable CFCs would slowly diffuse upward into the stratosphere, where intense solar ultraviolet radiation would cause the CFC molecules to release free chlorine atoms. These chlorine atoms react with ozone molecules and convert them to oxygen molecules according to the chemical reactions mentioned earlier.

The CFCs (chemical compounds containing chlorine, fluorine and carbon atoms) are chemically inert. They are non-toxic, non-flammable, invisible, tasteless and odourless. They replaced the toxic refrigerants sulphur dioxide and ammonia and this replacement saved many lives. Later, the inertness of CFCs led to their wide-spread use as propellants in spray cans and fire extinguishers and as a blowing agent in the production of plastic foams. In the 1950s, airconditioners began to be installed in cars and other vehicles (e.g., vans, used to carry food items). The CFCs used as refrigerants in these are much more likely to leak to the atmosphere than the CFCs used in household refrigerators and airconditioners. As a result, auto-airconditioners became a major source of CFC emissions.

There are many different CFCs; but the two most widely used CFCs are CFC-12 (CF_2Cl_2) and CFC-11 ($CFCl_3$), where the first digit in the name of the CFC shows the number of carbon atoms in it, while the second digit gives the number of fluorine atoms. Since the CFCs have hydrogen atom (H), they cannot be attacked by the atmospheric hydroxide (OH) radicals. The only process for removing the inert CFCs from the lower atmosphere is their slow transport to the top of the ozone layer in the stratosphere, where they are attacked by the solar ultraviolet light and release their chlorine atoms.

It has been discovered in recent years that, in addition to destroying the protective ozone layer in the stratosphere, the CFCs are major contributors to the Greenhouse Effect. The term **"Greenhouse Effect"** is used to indicate a heat-trapping process caused by certain "greenhouse" gases (such as carbon dioxide and water vapour), which are transparent to the incoming solar radiation, but absorb and re-emit the infrared radiation from the surface of the earth. What is causing a problem in recent years is the enormous rate of increase in the concentration of certain greenhouse gases in the atmosphere, leading to the possibility of global warming and large-scale climatic changes.

In addition to CFCs, some other gases can also attack the ozone layer. The examples of such gases are nitric oxide (NO) released by airplanes flying in the stratosphere and the relatively inert nitrogen dioxide (N_2O), if we release much of it at the ground level. NO, released by high-flying aircraft can contribute to the depletion of ozone layer by the reaction $NO + O_3 \rightarrow NO_2 + O_2$, which is swift and practically irreversible. On the other hand, since this reaction is not a catalytic one like the chlorine reaction, one NO molecule can destroy only one ozone (O_3) molecule; so much more NO is needed per unit of damage (to the protective ozone layer) than Cl.

8.4.5.4 *Protecting Ozone Layer*

So far the only method we know to protect the stratospheric ozone layer is to limit the emission of those substances (mainly CFCs) that can destroy it. No one knows of another material we could send into the stratosphere to protect the ozone layer from CFCs. The threat to the ozone layer is so severe that international conferences have been held and declarations and treaties have been adopted that commit the nations to restrict and eventually eliminate the use of CFCs. For some applications of the CFCs, satisfactory replacements are already available and for others, they are being sought on an emergency basis. Many of the proposed substitutes for CFCs are hydrochlorofluorocarbons (HCFCs), which contain atleast one hydrogen atom; so they are susceptible to attack by OH radicals in the stratosphere.

Huge amounts of money and effort are being spent now to develop suitable substitutes for the CFCs. The original CFCs were designed to be non-toxic, non-flammable and chemically inert. There replacements should have all these properties and, in addition, a low ozone depletion potential (ODP). The ODP is a relative measure of the ability of a gas or vapour to attack and destroy the stratospheric ozone layer. It is expressed as a function of the percentage weight of chlorine in the halogenated hydrocarbon (to be used a CFC substitute) and its lifetime in the stratosphere.

The chemical industry is developing two major types of CFC substitutes. One of these, known as the hydrochlorofluorocarbon (HCFC) family, has a fairly low ODP. The other group, known as the hydrofluorocarbon (HFC) family, has zero ozone depletion potential since the members of this family have no chlorine atoms in their molecules. It may be noted, however, that the members of both families (HCFC and HFC) contribute to the greenhouse effect and global warming. Many chemical companies, including the industrial giants like Du Pont (USA) and the Imperial Chemical Industries (UK), argue that the world should switch over to the HCFCs as an interim step because they are cheaper to produce and simpler to use.

Compared to the CFCs, the hydrofluorocarbon (HFCs) are less stable and readily hydrolysed in the troposphere by the OH attack; so they are believed to be more environmentally acceptable than the CFCs. The data on the new HFCs are, however, scarce and of limited accuracy and this is a major deterrent to their adoption as CFC substitutes.

Another problem with the CFC substitutes is that they are not as good as the CFCs. For example, one HCFC (known as HCFC-2), which is already being used in large-scale refrigerator installations in supermarkets, etc., cannot be used in domestic refrigerators. Another substitute, a hydrofluorocarbon known as HFC-134a, is under development for domestic refrigerators; but it is likely to cost at least five times more than the CFCs used currently for this purpose.

In addition to the problems mentioned above, the CFC substitutes will require a substantial amount of re-engineering in many applications. In case of domestic refrigerators, for example, larger compressors and pumps may be needed when CFC replacements are used. These substitutes may also require more energy, thus forcing us to burn larger amount of fossil fuel, leading to more global warming.

MODEL QUESTIONS

(Essay/Long Type)

8.1 Explain in brief the causes for environmental degradation in the Third World countries.

8.2 What is "organic growth"? Give some examples of unplanned growths.

8.3 Write a critical note on "Environmental Degradation by the Rich".

8.4 Explain in brief the effects of environmental degradation in India.

8.5 How is the ozone layer being destroyed? Explain in detail.

8.6 How should we protect the ozone layer? Write in brief.

8.7 Describe some methods to treat and dispose the municipal sewage.

(Objective/Short Type)

8.8 What are SO_x and NO_x?

8.9 List four heavy metals that pollute water.

8.10 What is a septic tank?

8.11 What are the e-wastes ?

8.12 What is ozone ?

8.13 The common methods for the disposal of solid wastes are :

    (a) Sanitary land-fill   (b) Incineration

    (c) Composting   (d) None of these.

8.14 India is a :

    (a) Developed country among developing nations

    (b) Developing country among developed nations

    (c) Both

    (d) Only (a).

8.15 Thermal power plants use :

    (a) Coal   (b) Oil

    (c) Uranium   (d) Strontium.

8.16 Purification of gaseous pollutants may be achieved by :

    (a) Catalytic conversion   (b) Adsorption

    (c) Absorption   (d) All of these.

8.17 Complex soil pollutants include :

    (a) Fly- ash   (b) Plastics

    (c) Both   (d) Neither.

8.18 Write true or false :

Herbicides, fungicides and insecticides belong to the family of pesticides.

8.19 Write true or false :

Detergents are inorganic chemicals.

8.20 Write true or false :

Radioactive isotopes have a very long half-life.

8.21 Write true or false :

Cadmium poisoning may lead to formation of kidney stones.

8.22 Write true or false :

Increase in dissolved oxygen adversely affects the aquatic plants and animals.

8.23 Write true or false :

Lime can not be used to precipitate phosphorous.

8.24 Write true or false :

The anaerobic treatment process is not effective for effluents containing high concentration of organic matters.

8.25 Fill up the blank :

Prolonged exposure to _____ may cause hearing loss.

8.26 Fill up the blank :

Gobar gas is a mixture of _____ and _____.

8.27 Fill up the blank :

Heat is an example of _____ pollutant.

8.28 Match the following :

(A) Noise pollution	(a) Carcinogin
(B) Dioxin	(b) Detritus tank
(C) Fly-ash	(c) Auditory effect
(D) Sewage	(d) Power plant

8.29 Match the following :

(A) Brain damage	(a) DDT
(B) Deforestation	(b) Population
(C) Third World countries	(d) Soil erosion
(D) White lung	(e) Textiles

8.30 Write the odd ones :

(a) Greenhouse effect

(b) Global warming

(c) Suspended particulate matters

(d) Acid rain.

9
Chapter

Waste Management

9.1 SOLID WASTES

9.1.1 Sources of Solid Wastes

Solid waste, the third pollution after air and water pollution, is that solid material which arises from various human activities and which is normally described as useless or unwanted. Solid waste consists of highly heterogeneous mass of discarded materials from the urban community as well as the more homogeneous accumulation of discarded materials from agricultural and industrial activities.

(1) Agricultural Wastes: These include wastes from feed lot, farm animal manure, crop residue, bagasse, etc. In India, agricultural wastes amount to roughly 2kg per person per day. Animal and vegetable wastes contain valuable minerals and nutrients. Humus from agricultural wastes contains nitrogen, phosphorus, potassium and many trace elements, which are vital to the fertility of the soil and optimum plant growth. Burning of agricultural wastes such as crop residue and cattle dung as domestic fuel makes poor use of the heat content of these wastes and further, leads to the loss of valuable nutrients.

(2) Urban Wastes: Domestic and commercial wastes are collectively known as **urban wastes.** The main constituents of urban wastes are very similar throughout the world; but the per capita waste generation, its density and the proportion of various constituents vary widely from country to country and from town to town within a country, according to the level of economic development, geographic location, weather and social conditions. As a general rule, it has been observed that as the family income rises, kitchen wastes decline; paper, glass and metals increase; the per capita waste generation rises and the density of the waste declines.

(3) Industrial Wastes: These are as diverse as the usable products generated by various industries. Industrial wastes include chemicals, paints, sand, scrap metals, fly-ash, sludge (from effluent treatment), etc. Manufacturing industries produce wastes which are solid or semi-solid. These wastes can be pyrophoric (i.e., self-igniting), explosive, toxic or even radioactive. Chemical process industries generate a variety of wastes, both organic and inorganic, which are mixtures of a wide range of chemicals in various proportions.

(4) Social Wastes: Social wastes may be classified partly on the basis of their contents and partly on their moisture and calorific values. A typical classification is:

(1) Garbage: This refers to putrescible (i.e., capable of decomposition by rotting) solid waste constituents produced during processing and storage of meat, fruits, vegetables, etc. Such wastes have a moisture content of approximately 70% and a calorific value of around 6×10^6 J/kg.

(2) Rubbish: This refers to non-putrescible solid waste constituents, either combustible or non-combustible. The moisture content of such wastes is around 25% and their calorific value is roughly 1.5×10^6 J/kg.

(3) Pathological Waste: This refers to dead animals, parts of human body (removed by surgery), etc. The moisture content of pathological waste is about 85%, while its calorific value is around 2.5×10^6 J/kg.

9.1.2 Public Health Aspects

The relationship between solid wastes and human diseases is difficult to prove. For example, if a flea, sustained by a rat, which in turn is sustained by an open garbage dump, bites a human being and transmits murine typhus, the absolute proof of the pathway would be to find the flea and the rat concerned, which is obviously not an easy task. Nevertheless, it is almost certain that improper handling and disposal of agricultural urban and industrial wastes is a true health hazard since we know that at least twenty-two human diseases are associated with these wastes. The two most important vectors (carriers) of human diseases in regard to solid wastes are files and rats. The fly is a prolific breeder (since more than two million flies can be produced in a week in one cubic meter of garbage) and a carrier of many diseases like bacillary dysentery. In India and many other developing countries with a warm climate, a common transmission route of bacillary dysentery, amoebic dysentery and diarrhea is from human faeces by flies, to food or water and then to humans. Rats, on the other hand, not only destroy property and infect by direct bite, but they are also dangerous as carriers of insects, which can finally act as vectors of various human diseases. The open refuse dumps also serve as a source of food for rats and other small rodents, which rapidly proliferate and spread to neighbouring areas. Rats, in addition to destroying property and infecting by direct bite, spread various diseases like plague, endemic typhus, salmonellosis and trichlnosis.

Apart from diseases for which flies and rats act as carriers, the handling and transportation of hospital wastes pose a serious threat to the workers handling these wastes and also to those who come in contact with them. Transmission of diseases may occur through direct contact with the wastes, through the infection of open cuts and sores, or through the carriers of various diseases.

Some of the industrial wastes are toxic and/or hazardous, and they pose a serious threat to human and animal health. These wastes have acute harmful effects, while others pose a health hazard after prolonged period of exposure. Improper handling and disposal of such hazardous or toxic wastes have resulted in the death of humans and animals through contamination of crops or water supplies. Moreover, there is the danger of water pollution when the leachate from a refuse dump enters surface or ground water resources. In addition, uncontrolled burning of wastes in open dumps can contribute to air pollution.

9.1.3 Collection of Solid Wastes

Efficient collection and transportation are essential parts of the overall solid waste management programme since these two activities account for about 75% of the total cost of solid waste management. The basic mode of refuse collection in our country is from various community storage points. The household refuse is delivered (usually on a daily basis) to fixed storage bins usually built from concrete blocks. These storage bins have a capacity of one to five cubic meters and are placed at intervals of 50 to 200 metres. The refuse is stored in the bins till it is collected by a vehicle (the garbage truck) for its final disposal. In case of urban solid waste, daily collection is essential because the organic matter in the refuse tends to decompose rather rapidly in a hot climate. The refuse is removed from the collection bins using a shovel and a basket and dumped into the waiting vehicle. Since the manual clearing of the bin is a slow process, the waiting time for the vehicle is excessive. Other methods of refuse collection such as block collection and curb side collection are practiced in developed countries. In the block collection method, the waste is brought in containers by the individuals to a waiting vehicle which travels a regular route, two or three times a week. The containers are emptied into the vehicle (by the vehicle crew) and then returned to the individuals. In the curb side collection method, the household waste is brought in containers and placed on the footpath in advance of the arrival time of the refuse collection vehicle. The empty container is retrieved later after departure of the collection vehicle.

Transportation of the refuse after its collection constitutes a key stage in the overall solid waste management programme. In India, no single mode of solid waste transport can be effective, economical and efficient due to our narrow and highly congested streets and lanes in most of our towns and cities. As a result, many different types of vehicles, ranging from handcarts to modern mechanized vehicles, are used for garbage collection in India. Handcart collection happens to be the best mode of refuse transport from our narrow and congested streets. The refuse is usually transferred from the carts to a waiting trailer for transport to the final disposal site. Large vehicles with some facility for mechanical compression of the refuse (to reduce its volume and to increase the density) are suitable for some of our well-planned cities, but smaller and simpler vehicles are better for most of the Indian towns and cities. Usually, the trailer or a truck is filled with refuse and then driven to the disposal site. If the disposal site is situated far away from the loading points, which is usually the case, the collection crew wastes a lot of time travelling to and from the disposal site.

9.1.4 Waste Disposal Methods

The problem of selecting the most appropriate method for the disposal of solid wastes is a complex one due to their heterogeneity; but proper selection can save money and avoid future problems for the municipality concerned. The disposal method should provide opportunities for recycling of useful materials, if possible and should not lead to pollution of air, surface and ground water and the land.

Several waste disposal methods are used at present in various parts of the world. The most prominent ones are:

(1) Open dumping.
(2) Sanitary land-filling.
(3) Incineration.
(4) Composting.

9.1.4.1 Open Dumping

Open dumping of agricultural, urban and industrial solid wastes is extensively practiced in India. The reason is that this method is very cheap and it does not require much planning. In general, the low-lying areas in the out-skirts of cities and towns are used. The open dumps cause public health problems by encouraging the breeding of rats, flies, mosquitoes, small rodents and other pests. Moreover, they also become a source of objectionable odours. When the wastes in the open dumps are burned (in order to reduce their volume and conserve space), they cause air pollution. Due to such drawbacks of open dumping, more acceptable methods of solid waste disposal are now used by most communities, usually at the insistence of state governments or Pollution Control Boards.

9.1.4.2 Sanitary Land-fills

Sanitary land-fills are engineered operations, which are designed and operated according to acceptable standards. The basic principle of a land-fill operation is to deposit the refuse, compact it with the help of bulldozers and then cover the material with at least 15cm of dirt at the conclusion of each day's operation. When the area is full, a final cover of about 60cm of dirt is applied, which is necessary to prevent rodents from burrowing into the refuse. The selection of a proper land-fill site is a difficult problem. The engineering aspects of land-fill site selection include:

(1) Drainage: Rapid runoff reduces the mosquito problems, but close proximity to streams or dug-wells may result in water pollution.

(2) Wind: It is preferable that the sanitary land-fill be situated in the down-wind direction from the city or town.

(3) Size: A small land-fill site with limited capacity is generally not acceptable because finding a new site involves considerable amount of trouble.

(4) Ultimate Use: At the planning stage, one has to consider whether the area can be utilised for public or private use after the land-filling operation is completed.

The above engineering problems are quite important, but even more important are the social and psychological problems of sanitary land-fills. No one in his right mind will be happy about having a sanitary land-fill near his house. A community may be rewarded with playgrounds, tennis courts, golf courses, etc., for tolerating a land-fill operation for a few years. If this operation is conducted according to accepted practice, there will be very little adverse environmental impact; but it is very difficult to convince and explain this to the people living near a land-fill. The reason is that most of the sanitary land-fills in the past were nothing but glorified garbage dumps.

The sanitary land-fill operation is, in fact, a biological method of waste treatment. In the absence of oxygen, anaerobic decomposition steadily degrades the organic material to more stable form. This process, however, is very slow. The decomposition may still be going on after 25 years. The end products of anaerobic decomposition are mostly gases like carbon dioxide

(CO_2), methane (CH_4), ammonia (NH_3), and a small amount of hydrogen sulphide (H_2S). Since these gases have to find an escape, it is a good practice to install vents in land-fills to prevent the build-up of the above gases. The decomposition reactions are self-sustaining and temperature inside the land-fill often reaches $55\text{-}60^0$ C.

The biological aspects of the land-fill as well as the structural properties of the compacted solid waste dictate the ultimate use of sanitary land-fill sites. It is suggested that nothing should be constructed on a land-fill site for at least two years because uneven setting may often create problems. If initial compaction is poor, it may take as long as five years for just 50% settling to occur. If pilings are used as foundations on a land-fill, they should extend through the fill and onto rock or some other sufficiently strong material.

9.1.4.3 Incineration

Incineration involves burning of solid wastes at high temperatures. After burning, the left-over materials (like ashes, glass and metals) amount to roughly 25% of the original wastes and they have to be disposed of in some manner. Incineration usually leads to air pollution unless the incineration plant is designed, equipped and operated to comply with air pollution standards. The main air pollutants resulting from incineration are fly-ash, sulphur dioxide (SO_2), hydrogen chloride (HCl) and some organic acids. Prior to incineration, materials which are not combustible are usually removed from the wastes by gravity or magnetic separation. Many of the separated materials like glass and metals can be recycled. Air pollution can be controlled by installing proper control equipment.

Industrial wastes that are incinerated are mostly cellulose-type wastes. More often, industries have to handle certain kind of chemical wastes in the form of either solid or sludge. In chemical process industries, incineration is most frequently used to dispose of tarry and gummy petroleum and plastic intermediate wastes and general refuse.

Combustibility of the waste is an important factor in determining the applicability of incineration. Waste combustibility is characterized by flammability limits, flash point and ignition temperature. The incinerator chosen must be capable of handling all types of solid wastes generated by the industrial plant concerned. Now-a-days, multiple hearth, rotary and fluidized-bed type incinerators are finding wide applications in industrial waste disposal.

If incineration is to become an economical method of solid waste disposal, then useful materials and energy (heat) must be recovered from this process. Heat can be recovered by putting a waste heat boiler or any other heat recovery device on an existing incinerator. The heat so recovered can be utilized for generating electricity or for the purpose of heating. In general, the solid waste has about one-third the calorific (i.e., heating) value of coal and unlike coal, it has a very low sulphur content.

The major advantages of incineration include the ability to handle a wide range of solid wastes and a small space requirement for the ultimate disposal of the left-overs after incineration. This method, however, requires a high level of maintenance and the operating costs are higher than those for other methods like sanitary land-fill.

9.1.4.4 Composting

In contrast to a sanitary land-fill which happens to be an anaerobic process, composting of solid wastes is an aerobic method of decomposing the wastes. Many types of micro-organisms, which are already present in the wastes, stabilise the organic matters in it to produce a soil conditioner. The micro-organisms that decompose the wastes include bacteria (which predominate at all stages of the process), fungi (which often appear after the first week) and actinomycetes (which exist during the final stages).

Initially, the process starts with the mesophilic bacteria, which oxidize the organic matters in the wastes to carbon dioxide (CO_2) and liberate heat. As a result, the temperature rises to about 45^0C. At this point, the thermophilic bacteria take over and continue the decomposition. During this phase of decomposition, the temperature rises further to about 60^0C. The refuse is periodically turned over to allow enough oxygen to penetrate all parts of the material to support the process. The compost is stabilized after about three weeks. The end point of a composting operation can be determined by noting a drop in the temperature, earthy smell and a dark brown colour.

The moisture content of the wastes is a critical factor in the aerobic composting process. Excessive moisture makes it difficult to maintain aerobic conditions. Dearth of moisture, on the other hand, inhibits the growth of micro-organisms. For an optimum rate of composting, a moisture content of about 55% is required, and it may be necessary to add water in order to maintain a satisfactory moisture content.

Modern methods of composting may be classified into two broad categories:

(1) Manual method.
(2) Mechanical method.

The manual method is more attractive for towns with a rural bias. Manual methods are widely practiced in India and many other developing countries. In India, two manual methods have been developed for composting the refuse/night-soil mixture. These are:

(1) The Indore process.
(2) The Bangalore process.

In the Indore process, layers of vegetable waste and night-soil are alternated, each about 8cm thick, to depth of about 1.5m in a trench, or form a mound above the ground. The mixture is kept aerobic by turning it at regular intervals for two or three months. The resulting compost is then left for another month or so without turning. The whole process takes about 3-4 months. The Bangalore process of composting is a modified form of the Indore process. The Bangalore process has been widely adopted now by various municipalities all over India. In this process, the refuse / night-soil mixture is placed in layers (as in the Indore process) in a trench of about one metre deep. In contrast to the Indore process, however, the material is not turned in the Bangalore process, but is digested under essentially anaerobic conditions, whereby the decomposition is complete in about 4-5 months. Though the Bangalore process avoids turning the material altogether, it lays great emphasis on the carbon to nitrogen (C/N) ratio of the compost heap and its initial moisture content.

The resulting compost is free from pathogenic (i.e., disease-causing) organisms. On dry basis, the compost contains approximately 1.5% nitrogen, 1.1% phosphorous (as P_2O_5) and 1.5% potassium (as K_2O) and proves to be a valuable nutrient for the soil.

Fully mechanized composting plants involve shredding, grinding and mechanical separation of high density solids. There are basically four processes of mechanical composting available in India:

(1) The Buhler process.
(2) The Dano process.
(3) The Tollemache process.
(4) The Nusoil process.

In the Buhler process, the waste material is ground in two stages in hammer mills. The non-compostable inorganic materials (metals and glass) are separated from the waste by the strong sifting action on circular swinging sieves. The remaining compostable material is then decomposed aerobically in open trenches or mounds. Stabilisation by this process may take about 2-3 months.

The Dano process uses a long rotating drum, called a bio-stabiliser unit, for decomposing the waste. The rotating drum is inclined so that the waste flows from one end to the other. The refuse is partially decomposed in the drum and the outcoming refuse is generally free from odour and pathogenic micro-organisms. This partially decomposed refuse is then completely decomposed in open trenches or mounds. The entire operation may take about a month.

In the Tollemache process, the waste is pulverized in a vertical pulveriser and then passed through a screening plant to screen out plastics, etc. The pulverised and screened refuse is allowed to decompose in trenches or mounds for about three weeks, with three or four turnings. The resulting compost is then cured for 4-5 weeks. The complete stabilization, therefore, takes about two months.

In the Nusoil process the non-compostable inorganic material is separated from the waste, the resultant waste is pulverised and the pulverized waste is led to a vertical digester where decomposition takes place. The vertical digester is a cylindrical unit having seven sections. The refuse moves downward through each section of the digester. The rates of air flow and water addition are regulated so that the decomposition of the waste takes place under optimum conditions. The refuse is kept for about a day in each section of the digester and thus, the digestion process is completed in about seven days. The resulting compost is quite satisfactory for direct field application without addition of supplementary nutrients.

9.2 MANAGEMENT OF AGRICULTURAL WASTES

9.2.1 Agricultural Wastes in India

Since India is a developing country, the Indian economy is very heavily dependent on agriculture. Roughly 70% of our population live in more than half a million villages and depend, directly or indirectly, on agriculture for their livelihood. As a result, huge quantities of agricultural wastes are generated each year. Given below is a brief list of agricultural wastes and their approximate generation, wherever known:

(1) Eighty species of non-edible oil seeds (1.0 Mt/year).
(2) Groundnut cake after the extraction of groundnut oil (2.5 Mt/year).
(3) Rice husk and rice straw.
(4) Tea waste (10,000 t/year).
(5) Bagasse (5.5 Mt/year).
(6) Molasses and press mud (0.2 Mt/Year).
(7) Tobacco waste (65,000 t/year).
(8) Wood waste, saw dust, bark from wood, etc.
(9) Coconut waste, including coconut shell (0.9 Mt/year), coconut husk (0.6 Mt/year) and coconut water.
(10) Lac Waste.
(11) Arecanut waste including arecanut husk (0.1 Mt/year) and arecanut leaf sheath (0.1 Mt/year).
(12) Cashewnut waste including cashew apple (30 Mt/year) and cashew testa (4,000 t/year).
(13) Coffee waste.
(14) Rubber waste including rubber seeds and rubber wood.

(15) Cotton waste including cotton stalks (25 Mt/year), cotton leaves and cotton linters.

(16) Jute waste including jute sticks (2.5 Mt/year) and jute mill waste (50,000 t/year).

(17) Fruit and vegetable waste including mango peels, mango kernels, citrus peels, citrus pomace, apple pomace, pineapple peels, banana stems and waste from vegetable-canning industries.

(18) Animal waste including cattle shed waste, fisheries waste, prawn waste, frog waste (6,000 t/year), animal carcasses (13 million per year), slaughter house waste and wool waste.

Further, we have **aquatic weeds,** which are indirectly related to agriculture through irrigation. Since aquatic weeds grow very prolifically, they block waterways and interfere with hydroelectric power generation. In addition, aquatic weeds waste valuable water by evapotranspiration (loss of water by its evaporation through the leaves of aquatic weeds), interfere with fishing activities and also help increase the spreading of water-born diseases.

The negative impact of aquatic weeds can be countered by converting them into food for fish, ducks, donkeys, pigs, sheep and cattle. They can also be utilized as raw materials to produce fertilizers, pulp, paper, fiber and energy (in the form of biogas). Water hyacinth, which is regarded as the most notorious aquatic weed, can be used for waste water treatment because of its property of rapidly absorbing heavy metals in the waste water.

In contrast to urban and industrial wastes, many of which can be hazardous, the agricultural wastes are non-hazardous in nature. They are, moreover, organic substances; so they are readily biodegradable. As a result, the management of agricultural wastes does not pose much of a problem. Agricultural wastes, in fact, can be regarded as a very valuable resource instead of a waste. If they are utilized properly and on a larger scale, agricultural wastes can contribute significantly to our growing economy. Instead of management, what we really need in connection with the agricultural wastes is their recycling on a massive scale to boost our economy. By recycling agricultural wastes, we can recover energy and many valuable materials.

9.2.2 Energy from Biomass

9.2.2.1 Biomass

Biomass is a general term used for all materials whose origin can be traced back to photosynthesis. Biomass includes all new plant growths, plant residues and wastes. Some important examples of biomass are wood, short-rotation trees, herbaceous plants, fresh-water algae, marine algae, aquatic plants, agricultural wastes, forest residues, crop residues (rice straw, rice husk, coconut and groundnut shells, bagasse, corn cobs, etc.), bark, sawdust, wood shavings, animal droppings and night soil. The positive environmental aspects of energy systems based on biomass are:

(1) Biomass is a renewable source of energy.
(2) Biomass acts as a sink for the atmospheric carbon dioxide since the plants consume carbon dioxide on photosynthesis and release oxygen. Plants, therefore, can control the adverse consequences of the greenhouse effect.
(3) Biomass helps to conserve soil and water since plants and trees reduce soil erosion and help enhance ground water resources.
(4) Biomass helps prevent the desertification of soil and halts the march of deserts.

Biomass in general and agricultural wastes in particular can be used as the fuel in thermal power plants to produce electricity; but this is not done in India since coal is readily available and relatively cheap. Some biomass-based energy systems for generating non-commercial forms of energy are described below:

9.2.2.2 Gasification

Gasification of biomass is one of the important routes to harvest energy through thermo-chemical conversion. It can yield biogas, producer gas and pyrogas. Agricultural wastes have calorific values comparable to other biomass like wood. Scientists at the Indian Institute of Science, Bangalore have developed a cyclone gasifier that can utilize agricultural and urban wastes for thermal or electrical power generation. Any combustible dry material (with a moisture content of 15% or less) in powdered form can be gasified to obtain producer gas in a cyclone gasifier. The gasifier can handle materials with high ash content and produce gas that can be used directly in thermal combustion engines. Scientists in Germany have devised a method to convert wood into a fuel gas, which can be used to generate heat and electricity. The process used in this case is **fluidized bed gasification**, which converts the wood into gas that is cooled and used as engine fuel.

Biomass gasifiers can convert agricultural wastes into clean and combustible producer gas that can be used to power agricultural and industrial equipment. The method consists of removing carbon monoxide (CO) and hydrogen sulphide (H_2S) from the crude producer gas, which is then allowed to undergo shift reactions with steam. In this process, the CO present in the crude gas gets converted into a hydrogen-rich gas that can be used in a fuel cell to produce electricity. Instead of industrial hydrogen, biomass can be used as a fuel-efficient and pollution-free system for rural electrification.

9.2.2.3 Biogas

It has been estimated that approximately 1,100 million tonnes of wet animal dung per annum are available in our country from 250 million cattles. Even at a 66% collection rate, this cattle dung can produce 22,500 million cubic metres of biogas through biogas plants and save 14,000 million liters of kerosene per year, which is mainly used in the villages for lighting and partly for cooking. In addition, the slurry from biogas plants can produce 210 million tonnes of organic manure, which can replace 1.4 million tonnes of nitrogen (N_2), 1.3 million tonnes of phosphate (P_2O_5) and 0.9 million tonnes of potash (K_2O) per year.

Unfortunately, this tremendous source of bioenergy has not yet been fully tapped. The variable factors range from cattle dung availability and its calorific value to biogas yield and appliance efficiency. Two major research needs in the area of biogas technology are restricted use of water in biogas plants and better strains of methane generating bacteria. In addition, temperature is a critical factor for the amount of methane generation in a biogas plant. The ideal temperature for maximum methane production is approximately 35^0 C since the bacteria are sensitive to lower temperatures, particularly in the northern regions of India. There is a fall in biogas production below 20^0 C and it stops altogether at 10^0 C. There is an urgent need, therefore, to develop new strains of methane generating bacteria which can produce methane at temperatures less than 20^0 C.

9.2.2.4 Briquetting

In India, approximately 350 million tonnes of agricultural wastes and 150 million tonnes of forestry wastes are produced every year. Unfortunately, the major part of these wastes is either not utilized at all or utilized very inefficiently. Scientists in India have developed a process, using the technique of pyrolysis, to convert agricultural and forestry wastes into charcoal briquettes that can benefit users and generate employment in rural India. In the pyrolysis process, the biomass is charred at temperatures ranging from 250 to 500^0 C for the partial removal of volatile substances. For this purpose, the pyrolyser (which is constructed from bricks and fine clay) is filled with biomass (i.e., agricultural wastes). Each pyrolysis operation with a charcoal yield of 100 kg takes about 6-8 hours. Charcoal obtained after the pyrolysis of wastes is crushed into fine powder. After grinding, a binder (about 10% by weight) is added to charcoal power. In order to increase agglomeration of this mixture, about 6% clay (by weight) is added to it. The final mixture of the charcoal powder, binder and clay is then converted into briquettes by applying pressure.

Scientists at the Indian Institute of Technology, New Delhi have developed and perfected the technology for briquetting saw dust, plant leaves, crop residues and other similar wastes into a smokeless fuel. This technology is now being promoted through pilot demonstration projects in various parts of the country.

9.2.3 Agricultural Waste Recycling

Various methods of conserving our resources (both materials and energy) are generally known as recycling. In recycling, a product at the end of its service life is converted into another useful product or raw material for some useful product. The process of recycling involves separating, collecting, processing, marketing and ultimately using the material that would have been thrown away as waste. When a product at the end of its service life has been recycled and then used as a new product, the recycling loop is closed. The process of creating a new product from the waste is extremely efficient since it produces virtually no waste. The most important advantages of waste recycling are:

(1) Recycling of waste reduces dependence on land-fills and incinerators for waste disposal.
(2) It protects health and environment by effective handling of agricultural, urban and industrial wastes.
(3) Recycling conserves limited resources because it reduces the need for fresh raw materials.

Today we can recycle most of the materials (such as metals, paper, synthetics, glass, plastics, etc.), which are found in our wastes. For stable economic growth and sustainable development, material and energy resources must be used carefully and technologies for waste recycling must be evolved. We have to make a choice between one-time use of materials (the throw-away society) and use of recycled materials (the sustainable society). The latter choice will make scarce materials available to future generations and will also save our environment.

In India, huge amounts of agricultural wastes are readily available. Only a small fraction of these wastes is recycled. There is an urgent need to recycle our agricultural and other wastes on a much larger scale in order to accelerate our economic growth. An agricultural waste, sugarcane bagasse, for example, is an important source of cellulose; but it is burnt away now as a cheap fuel. For better economy, it is important to get some useful byproducts from bagasse. Now-a-days, agricultural waste is increasingly being used as a raw material for the manufacture of reconstituted panels as a near substitute to timber-based panels. Development of medium-density fibre has been started in the US. In this process, fibres processed from crop stalks are converted into high-quality panels, which are almost equivalent to wood. This process, therefore, has the potential to save our dwindling forests. Several important processes for the recycling of agricultural waste are described below:

266

9.2.3.1 Cellulose from Groundnut Shells

Microcrystalline cellulose (known more commonly as MCC) provides many of our daily requirements and comforts such as clothes, food and paper. MCC has also been widely used as a filler material in the food processing industry. Cellulose can be obtained from a number of sources such as bamboo, wheat stalk, sugarcane bagasse, cereals and waste paper. Groundnut is one of the major agricultural crops grown in our country. Groundnut shells have been found to contain about 60% of crude fibre.

A new method for obtaining MCC from waste groundnut shells has been developed in our country on a pilot plant scale. This process consists of refining, cellulose pulping, bleaching of the pulp and hydrolysis. In the optimized pilot plant process, the groundnut shells are first sieved and then treated with nitric acid in a glass-lined, steam-jacketed reactor. After the nitric acid treatment and washing with water, the material is immediately led to a stainless steel reactor, where it is treated with sodium hydroxide (NaOH). The material is then washed and the product is subjected to bleaching process followed by drying and pulverization. The result is 98.53% pure microcrystalline cellulose of pharmaceutical grade. The final product is white, crystalline, odourless, tasteless and fine in nature.

9.2.3.2 Particle Board from Rice Husk

Scientists at the Indian Plywood Research Institute, Bangalore have developed a technology for manufacturing particle board from rice husk. The particle board obtained by using this technology is a versatile substitute in a wide range of applications. The particle boards, moreover, can also be made decorative by incorporating suitable colours; so they can be made more elegant looking and attractive than any other wood or plywood substitute. The particle board has been found to be superior to wood or plywood boards because of the following advantages it offers:

(1) Particle board is resistant to termites (white ants).
(2) It has a high resistance against decay.
(3) The particle board has excellent mechanical properties such as elasticity, dimensional stability, screw and nail holding capacity, abrasion resistance and surface hardness.
(4) It has an improved water resistance.
(5) It has a high durability.

Depending on the market demand, a variety of product-mix can be produced by varying the density and resin content of the boards. The method of particle board manufacture involves the following basic steps:

(1) Cleaning of the raw material (the rice husk).

(2) Application of adhesive.

(3) Mat forming.

(4) Edge trimming.

(5) Sanding.

(6) Stacking.

The process has been licenced to several firms in India and Malaysia. The particle boards produced by these firms have shown excellent market acceptance and gained popularity because of their elegant look, better mechanical properties and lower prices.

9.2.3.3 *Paper and Board from Jute Wastes*

Researchers at the Jute Technological Research Institute, Kolkata have developed the technology to produce different grades of special papers and boards using jute fibres and jute wastes as raw materials. Jute fibres in the form of jute caddies, root cuttings and thread wastes, which contain a lot of cellulose, can serve as non-wood sources for making different grades of paper. These include wrapping paper, sack paper, paper board and writing paper. The technology can help reduce the use of precious wood used in paper making.

The socio-economic significance of jute cultivation lies in the fact that about 10 million small and marginal farmers in India, Bangaladesh, China, Nepal and Thailand depend on jute for their livelihood. Jute, moreover, is an environment-friendly fibre when compared to artificial fibres like poly propylene (PP).

After harvesting, jute plants are defoliated in the field itself. Thereafter, jute stems are retted (i.e., softened) in water and the fibre is extracted. The waste of jute production is the dry matter in the form of jute sticks that cannot be used as fibre. Each tonne of dry jute fibre generates approximately 4.5 tonnes of dry jute leaves and sticks. Dry leaves can be used as manure and jute sticks as construction material or fuel. Jute products, jute-based paper and boards and jute composites can directly help protect the environment. If more jute is used for paper making, more forests can be saved.

9.2.3.4 *Medicines from Agricultural Waste*

Furan compounds occur widely in nature and they are cheap raw materials. The compound furfural, for example, is readily obtainable from agricultural wastes such as corn cobs and oat hulls. Furfural, which is commercially produced by the reaction of corn cobs with sulphuric acid, is the basic material used for synthesis of nitrofurans, which are important germicides used for treating cattle diseases. Nitrofurans are also used to control the bacteria commonly found in the intestine and causing diseases such as gastro-enteritis, typhoid, diarrhea and gastro-intestinal haemorrhage. Furfural is also obtainable from sugarcane bagasse.

9.3 MANAGEMENT OF URBAN WASTES

9.3.1 Composition of Urban Wastes

Domestic and commercial wastes together are known as urban wastes. The major constituents of urban wastes are similar every where but their quantities and proportions may vary widely. In our own country, authentic information regarding the composition of urban wastes is not generally available since regular analysis of the refuse is rarely carried out by municipalities. Moreover, urban wastes are very heterogeneous in composition. Geographical, temporal and seasonal variations in the composition of urban wastes make it difficult to define a "typical refuse". Solid wastes generated in urban areas contain items of various types and sizes such as vegetable peels and leaves, fruit peels and kernels, waste paper, cardboard cartons, glass bottles, metal containers, carcasses of animals, coconut shells and husk, pieces of stones and bricks, dust, etc. Studies on the composition of urban wastes have been conducted in major cities like New Delhi, Mumbai, Kolkata, Chennai, Kanpur and Bangalore and even in a small city like Rourkela. Based on these studies, the percentage (by weight) of various constituents in the urban wastes is found to be as follows:

(1) Paper (0.14 – 5.88%).
(2) Vegetable (putrescible) residues (47.25 – 75.2%).
(3) Ash, dust, etc. (12.00 – 33.58%).
(4) Metals (0.10 – 0.66%).
(5) Glass (0.20 – 0.52%).
(6) Textiles (0.28 – 3.56%).
(7) Plastics (0.66 – 1.54%).
(8) Other items such as stones, wooden matter, etc. (6.40 – 18.90%).

The density of urban wastes in our country, which varies from 540 to 578 kg/m^3, is much higher than that of the wastes generated in Western countries (roughly 132 kg/m^3) because the Indian urban wastes include street sweepings in it. The amount of refuse collected from urban areas in India is of the order of 0.50 – 1.5kg per person per day (excluding night soil). Efficient and economical management of urban wastes involve the following steps:

(1) Collection of waste.
(2) Volume reduction.
(3) Resource recovery.

(4) Waste recycling.
(5) Waste disposal.

Among these, the first and the last step, viz., collection of waste and waste disposal (through open dumping, sanitary land-fill, incineration and composting) have already been discussed in Sec. 9.1. The remaining three steps are discussed in the following sections:

9.3.2 Volume Reduction

Urban waste is a bulky material which does not compact easily. On the other hand, the volume requirements in a sanitary land-fill are quite significant since land is expensive and the cost of land-filling can be high. To solve this problem, various methods of reducing the volume of refuse prior to its disposal have been devised.

Incineration has been often touted as the "final solution" to the solid waste problem (including volume reduction). However, incineration only reduces the volume of solid waste to 20 – 30% of the original volume and makes the product stable; but the residue left after incineration has still to be disposed of in some manner. Incinerators, moreover, have high capital costs. Similarly, the operating costs of incinerators can be quite significant. The cost of incineration is increased due to the requirements of air pollution control. Another problem with incineration of urban wastes is that the power supply in most Indian cities and towns is erratic and unreliable; so efficient operation of incinerators becomes a difficult task. When the voltage is low, the required high temperature is not reached. As a result, the problem of air pollution is aggravated due to incomplete combustion of waste.

Shredding the solid wastes (also known as *pulverizing*) and then spreading the shredded material on fields has been found to be successful in a number of places. Since spreading it out dries the pulverized refuse, it avoids the problems of offensive rats and files. The shredded material does not have to be covered with dirt, which is a significant advantage over the sanitary land-fill. The reason is that the capacities of many land-fill sites are limited not by the available volume but by the availability of dirt to cover the refuse. In addition, land-fill operations are quite difficult to carry out during the rainy season.

Pyrolysis is the combustion of urban waste in the absence of oxygen. The main advantage of pyrolysis over incineration for volume reduction is that the residues have economic value. On the other hand, the residues of pyrolysis, viz., combustible gas tar and charcoal, have not yet found ready acceptance as raw materials. Most of the combustible gases, in fact, is used in processing the process of pyrolysis. The tar obtained from pyrolysis is full of water and it has to be refined. The charcoal, on the other hand, is full of glass and metal, which must be separated before it can be used. Other advantages of pyrolysis are that it reduces the volume of solid wastes considerably, produces a stable end product and it has fewer air pollution problems compared to incineration (due to smaller volume of gas to be treated). On a larger scale (in the case of larger cities and towns), pyrolysis as a method of volume reduction has significant advantages over incineration of urban wastes. However, the value of pyrolysis as a treatment method for materials recovery is yet to be demonstrated.

9.3.3 Resource Recovery

9.3.3.1 Economic Considerations

Urban wastes contain significant amount of valuable materials like steel, aluminium, copper, other metals, paper, glass, and plastics. These materials, if recovered, would reduce the volume of the waste to be collected and disposed of and, at the same time, would yield significant salvage and resale value. Better reclamation techniques would help to save valuable natural resources and turn wastes, which could be dangerous for health and environment, into useful products. Recovery of resources from wastes is theoretically a very appealing concept. Unfortunately, our present economic system, which is primarily geared to profit making, makes recycling of waste difficult. The picture is changing rapidly and in the near future, resource recovery may become the most desirable means of waste management.

Not much attention has been paid to the **total process** of resource recovery. Resource recovery or recycling has often been confused with the mere collection of a potentially useful material like old newspapers and glass bottles; but collection is only the first step in the total process of recycling. After the material is collected, it must be cleaned, sold to an industry as a raw material, converted into some useful product by the industry concerned and, most importantly, sold once again to the consumers. The last step, in fact, controls the entire operation. If the final product manufactured from recycled raw materials can not be sold at a profit, there is little economic sense in carrying out other operations of the recycling process.

The two basic reasons for resource recovery are:

(1) Conservation of resources.
(2) Reduction in volume of the refuse to be disposed.

Some of the important materials that have been successfully reclaimed from urban and industrial wastes are paper, glass, metals, plastics and organic materials.

9.3.3.2 Paper

The paper industry treats the incoming waste paper according to its sources. Paper recovered from the urban waste goes directly to the paper mills. Waste paper contains grit, sand, ink, tar, paper clips and pins, plastic coatings (as lamination) and rubber bands, which have to be separated before the waste paper can be treated further. Only a small fraction of recycled paper ends up as virgin paper. More commonly, the used and recycled paper is converted to paper board, insulation, paper bags, packaging paper, etc. Recycled paper can be used for printing only a few times because, with each operation of recycling, the paper fibres become somewhat shorter and more frayed, with the result that the recycled product is weaker than a similar product made from virgin fibres.

The manufacture of paper board is the main use of waste paper. After sorting into different grades, the waste paper is fed on to a conveyor belt in the proportion required by the mix. The conveyor belt feeds the waste paper to a hydropulper unit, where it is repulped in water, with a mixture of 3 parts of paper to 97 parts of water, at about 38°C. The stirring force in the hydropulper is so large that the sheets of waste paper are rapidly broken down to a slurry. The fibres in the paper are separated and heavy, unacceptable materials are drained from the bottom. The slushed waste paper is then passed over a riffler system and light material is floated off. The slush is screened to remove any undissolved matter and it is then fed to a bitumen dispersal unit for removal of tar. The slushed paper slurry is then thickened and the stock is ready for the paper board machines.

Perhaps the most promising use (or reuse) of waste paper is in its conversion to energy. Waste paper has a high calorific value (approximately half of that of coal) and it does not cause air pollution like coal. The plastic and wax coatings enhance the value of waste paper as a fuel and contribute to its heating value.

9.3.3.3 *Glass*

Glass is the most perfect product for recycling. It is clean, easy to reprocess, and it can be used in many ways. At present, ready markets exist in our country for waste glass. Scrap glass can be used in each new batch of glass produced. The main problem, however, in any reuse of glass is its separation from other materials. The typical glass separation methods include froth flotation, dense media separation and colour sorting.

Waste glass can also be used as an aggregate in asphalt instead of crushed stones. The finished product exposes no jagged edges and is skid resistant. Several industries in the US have experimented with a water-soluble glass that can be treated with a thin protective coating to keep the glass from dissolving until the container is broken. Experimental use of ground glass as an ingredient of light-weight building materials has also been technically successful.

9.3.3.4 Metals (Re: 10.5.2.)

9.3.3.5 Plastics

Plastics form a small percentage of the total solid wastes in India, but their use is increasing at an accelerated rate. Plastics basically consist of carbon, hydrogen and oxygen; but they are not biodegradable. As a result, they remain intact in waste disposal operations. Polyolefins, polystyrene and polyvinylchloride (PVC) make up more than 80% of the total quantity of plastics in the urban wastes.

One possible method of recycling plastic is to use it in the form in which it was first manufactured, i.e., by mixing the raw plastic during the production of new packs. There are problems, however, due to a large number of different grades of materials within each major type of plastics and the grade employed depends on the characteristics required in a particular pack.

Since mixed plastics cannot be recycled in the manner mentioned above and mixed plastics are the most common form of plastics in the urban waste, recycling has to be done by reusing the plastic recovered from urban wastes in some other way. The Union Carbide Company in the US has pyrolysed plastics to produce hard and soft wax, grease, adhesive and tar as well as a range of useful gases.

Another idea for the recycling of plastics, which is being carefully investigated, is the total decomposition of plastic materials. When exposed to ultraviolet light, certain chemical groups absorb the ultraviolet light and emit energy. This energy breaks the polymer chain so that the plastic looses its strength, becomes brittle, and is broken up by the wind and rain. The small pieces of broken plastics then mix with the soil and can be decomposed by normal soil bacteria into basic chemicals.

9.3.3.6 Organic Materials

A very large portion of the urban wastes consists of organic materials, which can be converted into several useful products as well as energy. The most common process, used extensively in many countries, is composting.

A composting operation usually involves the following steps:

(1) Segregation of the urban waste into organic and inorganic components.
(2) Grinding of the organic portion.
(3) Stabilising it in either open piles or in mechanical digesters.

The segregation operation is the weakest link in the process of composting. Most municipalities have not had much luck in asking the citizens to separate paper and garbage from glass bottles, tin cans and plastic items. Manual sorting at the composting plant, on the other hand, is an expensive and time-consuming process. In our country, however, rag-pickers often do part of this sorting. Considerable research is being directed now towards finding better separation methods. It is felt that if this problem were solved, composting would be a much more attractive method of resource recovery from urban wastes.

Proper handling and recycling of urban wastes can provide energy and lower the waste burden on the natural environment. In the following section, a few methods to obtain useful products by recycling urban wastes are described in brief:

9.3.4 Urban Waste Recycling

9.3.4.1 Fuel from Plastic Waste

The center for Microporous Materials in Manchester (UK) has developed a process for distillation of fuel from dirty, mixed plastics recovered from urban wastes. The Center has developed zeolites, the crystalline aluminium silicates, for producing the fuel from plastic wastes. The zeolites act as catalysts in the distillation of fuel. Since the fuel obtained from waste plastic contains no lead and has a high octane rating, it is known as "**green fuel**". This fuel will not only help in reducing the use of petrol, but it will also help solve the environmental problem of disposing of the non-biodegradable plastic waste.

The process of distillation involves heating of plastic waste in an oxygen-free environment in order to prevent the burning of plastics. The hydrocarbons in the plastics are distilled of as vapours, which are cooled and collected. A significant portion of the plastic waste consists of PVC (polyvinylchloride), which produces hydrochloric acid (HCl) when heated. The acid can be broken down into a harmless compound by adding an appropriate zeolite catalyst during distillation.

A waste recycling plant in Germany has developed a technology for converting assorted plastic wastes into oil. The plastic wastes such as polythene bags, milk pouches, plastic cartons, computer casings, etc., are mixed with heavier oil residues and then water is added. As a result, chemical compounds in the waste plastic break up to produce synthetic oil, which is piped to a refinery to yield high-quality oil-based products. In some plants, instead of adding water to it, the plastic waste is heated after mixing it with heavier oil residues. By this process, one kg of waste plastic yields about 0.9kg of reusable raw materials.

9.3.4.2 Recycled Plastic Clothes

Clothes from recycled plastic waste will soon be available in the global textile market. Going by the recent developments in the chemical and textile industries, it may not be long before a sweater or a jacket made from recycled plastic bottles becomes an essential fashion accessory.

The Brasher Boot Company, a small British concern, launched Europe's first fleece jacket in London, which was made from recycled polyethylene terephthalate (PET)- a plastic used in making soft drink bottles. Except for the zipper and the thread, the rest of the jacket was made from recycled PET. This Company turned to recycled plastic fleece on commercial as well as environmental grounds, and recycled materials are forming an increasing part of the Company's output. Garments made from recycled PET meet all the basic demands of a clothing material, viz., lightness, comfort, warmth and durability, in addition to being environment friendly.

Rhovyl, a company in France, has started making yarn comprising 70% polyvinylchloride (PVC) derived from the recycling of mineral water bottles. The yarn is as soft and supple as the natural wool.

9.3.4.3 Power Generation from Garbage

Our cities and towns are cursed with piles of rotting garbage and endless hours of power cut. Western Union India is now offering a solution to both the problems simultaneously. The multi-national company has developed a technology to produce electricity from city garbage. A Rs.100-crore plant can process 1,000 tonnes of city garbage per day and generate 8 MW (megawatts) of electricity, making a profit of Rs.25 – Rs.30 crores per year. In India, the company has entered into an agreement with several municipal corporations, including Kolkata, Aurangabad and Thiruvananthapuram. Earlier, it had signed agreements with the governments of Vietnam, Turkey and Thailand.

In the technology developed by the Western Union India, city garbage is used to produce methane and the latter can be used to generate electricity. The technology is ideal for the Indian garbage, which has a high concentration of biodegradable matter and a high moisture content, unlike the garbage produced in Western countries, which has a high concentration of combustible materials like paper. The company has already set up a pilot plant at Pune, which can generate 5kW of electricity. At this plant, ferrous metals are removed from the garbage by using a powerful magnet. Paper, sand, plastics, glass, etc., are removed by using a pneumatic separator. The garbage is then mixed with water and bacterial species, specifically cultured for producing methane. These bacteria thrive on the waste and produce methane by digesting the refuse. This process is known as **biomethanation.**

The bacteria take about a day for the conversion of garbage into methane. The reactor used for this conversion is a double-reinforced cement concrete structure, which is usually maintained at a temperature of about 15^0 C. As a result, about 65% of the garbage get converted into methane and the remaining product is used as manure. The methane so produced is then fed into a gas-fueled internal combustion engine, which runs a generator to produce electricity. It is claimed that 500 MW of power can be generated in India from our urban solid wastes alone and about 2,000 MW of power can be generated by using both municipal solid wastes and industrial effluents.

9.4 MANAGEMENT OF INDUSTRIAL WASTES

9.4.1 Industrial Wastes

Manufacturing industries produce wastes which may be gaseous or non-gaseous. In terms of physical states, the non-gaseous industrial wastes may be classified as follows:

(1) Liquids.
(2) Slurries.
(3) Sludges.
(4) Solids.

Industrial wastes may also be classified as: (1) hazardous and (2) non-hazardous. A large proportion of industrial wastes are non-hazardous. The two principal methods used for the disposal of non-hazardous industrial wastes are sanitary land-fill method and incineration method. Certain industrial wastes may also be disposed of by special burial. Composting is rarely applicable to industrial wastes. There is a considerable scope of recovery and recycling of useful materials in the case of industrial wastes. Most of the industrial wastes generated in cities and towns come from small-scale industries. They are, in general, non-hazardous; so they are usually disposed of along with the urban waste. Larger industries are often located outside the cities and the disposal of their wastes is primarily the responsibility of the industries themselves. Some of the industrial wastes like scrap metal and paper can be recycled, while others can be utilised as a source of energy for specific processing plants. Energy can be recovered from industrial and other solid wastes by numerous thermal routes as well as by biochemical conversion (in composting, for example).

Hazardous industrial wastes need special consideration before their disposal. They include the following categories of industrial wastes:

(1) Explosives.
(2) Compressed gases.
(3) Liquefied gases.
(4) Gases dissolved under pressure.
(5) Inflammable liquids.
(6) Inflammable solids.
(7) Inflammable substances liable to spontaneous combustion.
(8) Inflammable substances which emit inflammable gases when in contact with water.

(9) Oxidising substances.

(10) Organic peroxides.

(11) Toxic poisonous substances.

(12) Radioactive substances.

(13) Corrosives.

(14) Other dangerous chemicals in limited quantities.

Industrial wastes are as diverse as the usable products generated. An industry with solid wastes has to choose among the following courses of action:

(1) To reprocess the wastes into new products, e.g., using fly-ash to make bricks.

(2) To dispose of the wastes.

(3) To store the wastes until better ideas come along.

Whatever method is adopted for the disposal of industrial wastes, it may be necessary for the wastes to be segregated, collected, transported and possibly processed prior to their final disposal. Whenever reasonably possible, industrial wastes of different classes (hazardous and non-hazardous) or of different contents (solids, liquids, slurries or sludges) should be segregated to allow for separate collection, transportation and disposal.

9.4.2 Disposal of Industrial Wastes

9.4.2.1 Land-fill

In the first place, some industrial wastes pose special problems for land disposal. Such wastes may need to be processed before their disposal in a land-fill. An example of such problematic wastes is oily sludge, which cannot be satisfactorily buried because it persists for years almost unchanged and it can migrate. Industrial wastes which are difficult to dispose of safely on land include the following:

(1) Non-biodegradable hazardous materials.

(2) Chlorinated hydrocarbons.

(3) Slowly degradable materials.

(4) Mineral oils and greases.

(5) Persistent materials (heavy metals like chromium, cadmium, lead mercury, nickel, copper and zinc).

(6) Toxic (poisonous) substances such as phenols, cyanides (including metalcyanide complexes), drug residues, pesticide residues, etc.

(7) Strong acid or alkaline materials in quantities which would disturb the neutrality of the land-fill.

(8) Acids (hydrochloric, sulphuric, nitric, phosphoric, chromic or hydrofluoric).

(9) Sodium hydroxide.

(10) Non-aqueous liquid wastes which would dissolve materials used in sealing a land-fill site.

No land-fill site should be used for the disposal of industrial wastes which include the materials listed above unless the permeability of the ground is such that leakages to underground water supplies can not occur within at least 250 years. Sites selected for land-fill are typically below the level of the surrounding land, either natural depressions or excavations. Clay pits are useful since the excavations there have impervious linings. In some instances (on sites located on estuaries), artificial mounds of wastes are created to build the land up above its original level. In any case, disposal of toxic wastes by land-fill should always involve adoption of techniques for rapid covering of the deposited material (if solid) or for safe absorption of the waste (if liquid) into previously deposited innocuous wastes. Soakage trenches are used for this purpose.

9.4.2.2 Other Processes

Besides land-fill, the other principal method used for the disposal of industrial solid wastes is incineration which has already been discussed earlier. A process is now available to convert *hazardous wastes* in solid, liquid or sludge form into a hard polymer. The hazardous industrial wastes are conveyed to a series of containers fitted with high-speed disintegrators for the purpose of dissolution and dispersion. Chemical pre-treatment may be applied to some of these wastes. Polymerisation is limited by the addition of special reagents to produce the polymer in slurry form. This is pumped to an adjacent disused clay pit, where it sets in three days to give a hard, rock-like solid. In this process, no gases or liquid effluents are produced. A wide variety of wastes (excluding the combustible or inflammable ones) can be treated by this method. Polymerised wastes have low permeability, good resistance to leaching by water and good physical strength; but they are not biodegradable.

9.4.2.3 Effluent Disposal to Tidal Water

Many considerations arise in the satisfactory use of tidal water for the disposal of industrial effluents (or municipal sewage) in the water or on the shore. We have to consider, the conservation of marine flora and fauna, commercial navigation and fishing. Moreover, the disposal system must not be subject to damage by shipping or weather. Consideration has to be given also to the possible long-term effects of silting and erosion. The tidal water outfall should be designed to give the required primary dilution to the effluents at the outfall. A suitable diffuser device may be provided for this purpose. Location of the outfall should give adequate further dispersion to effluents to allow the concentration of any persistent materials to approach the background levels and to give time for their natural treatment. A study of the hydrology of the sea area is required in selection of the outfall position for good secondary dilution, so as to make the full use of tidal movement.

9.4.2.4 *Underground Disposal*

In addition to surface tipping in old quarries or other land-fill sites, industrial solid wastes may also be disposed of in boreholes and disused mineshafts. Provided they are carefully selected, mineshafts may shield the dumped waste material from contact with circulating water, thus protecting the environment from contamination. Pumping of liquid wastes under pressure into porous and permeable rocks at depth is widely practiced in the US and some other countries. However, many industrial wastes disposed of in this way irreversibly pollute the rocks and may, therefore, form an environmental hazard. The use of old quarries may endanger ground water sources where the enclosing rocks are porous as well as permeable. Selection of underground disposal sites requires careful consideration of the hydrogeological conditions in the area and stringent controls must be exercised. In the same way, when a site is considered for surface disposal of industrial wastes, both long-term and short-term risks of environmental pollution must be assessed. Monitoring controls must be developed to detect and asses the movement of effluents in the subsurface in time, so that the pollution risks are minimized and potable water sources (both present and future) are protected.

9.4.3 Resource Recovery

9.4.3.1 *Recovery of Metals (Re: 10.5.2.)*

9.4.3.2 *Materials Recovery from Liquid Wastes*

Considerable quantities of metals are lost annually in industrial effluents, effluent treatment sludges and filter cakes. Attempts have been made to recover valuable metallic constituents from effluents by using hydrometallurgical techniques. Because of the relatively higher value and the ease of recovery of metals, there have been many proposals for the reclamation of metals from specific waste solutions. Some examples of these proposals are:

(1) Recovery of silver and other precious metals from electroplating rinse water.
(2) Copper compounds from printed circuit board.
(3) Iron oxides from spent pickle liquor.

Most of the processes used for metal recovery from industrial effluents are based on hydrometallurgy. Many other valuable materials can be recovered from industrial effluents and solid wastes. Some examples of such recoverable materials along with their source (mentioned in parentheses) are as follows:

(1) Acetone and carboxylic acids (petrochemicals).

(2) Ammonia, ammonium sulphate, aromatic organic substances, naphthalene, tar, toluene and xylenes (coke oven wastes).

(3) Anthracene, creosote and cresol (coal tar).

(4) Anthranilic acid, methylaniline, potassium hydroxide and sodium hydroxide (dyestuffs wastes).

(5) Aromatic substances (naphtha pyrolysis quench).

(6) Benzene and methylbenzene (styrene plant effluents).

(7) Chromic acid, gold cyanide, nickel salts and sliver cyanide (electroplating wastes).

(8) Ethanol (cellulose wastes).

(9) Ethylene dichloride (oxychlorination waste water).

(10) Hydrochloric acid, iron sulphate and sulphuric acid (pickle liquor).

(11) Hydrogen and phenols (refineries wastes).

(12) Lubricating oil (wastes from vehicle maintenance units).

(13) Paint (automotive plant wastes).

(14) Protein (fermentation effluents).

(15) Quench oil (metal treatment wastes).

(16) Sodium chloride (desalination plant wastes).

(17) Sodium sulphate (paper mill effluents).

(18) Sulphur (wastes from desulphurisation units).

(19) Sulphur oxides (ore smelting wastes).

(20) Trichlorobenzene (pigments unit wastes).

(21) Trichloroethylene (degreasing wastes).

(22) Yeast (Brewery wastes).

(23) Zinc salts (rayon plant wastes).

The above list illustrates the diversity of recyclable materials that can be recovered from industrial liquid wastes. In the following section, we describe how some very useful products can be manufactured from industrial wastes:

9.4.4 Industrial Waste Recycling

9.4.4.1 Wealth from Fly-ash

Fly-ash from coal-fired power plants is a major industrial waste product in India. Since our country is surplus in coal, we have many coal-based thermal and super-thermal power plants in operation, and many more are in the process of being installed. Annual availability of fly-ash in India is 25-30 million tonnes. Different uses to which fly-ash is put at present are as follows:

(1) Cement and concrete (including light-weight cement, foam light-weight cement).
(2) Land and land foundation (including land development, fragile ground improvement materials, asphalt filters, fill-type dam materials, blocks for construction).
(3) Agriculture and fisheries (including soil improvement, potash fertilizers, gathering ground for fish, compost for night soil treatment).
(4) Building materials (including rock wool, glass wool, etc.).
(5) Sewer treatment (including deodorants, sludge treatment materials, sewer treatment materials).

Due to shortage of oil in India and abundance of coal, there is an increasing shift to coal as a source of energy. Moreover, the use of fly-ash is also gathering ground on account of the fact that mineral matter in the fly-ash has already been heat-treated during the burning of coal. Obviously, less heat is needed to fire (or heat-harden) bricks made from the fly-ash. Various advantages of bricks prepared from the fly-ash are as follows:

(1) These are fired at lower temperatures because the fly-ash has already been heat-treated earlier. This gives one-third of saving in the cost of bricks.
(2) Bricks using fly-ash take less time to dry and fire; and they require simpler kilns.
(3) Fly-ash is cheaper than clay.
(4) Coal being a major source of power in India and many other developing countries, its waste product (fly-ash) can be gainfully utilized, which otherwise could be an environmental hazard.
(5) Fly-ash can also be used beneficially in the pottery industry.

9.4.4.2 Cash from Trash

Tata Chemicals has set up plants to recover valuable chemicals like magnesium, soda ash and calcium carbonate from the effluents discharged by its various plants in India. Recovered magnesium is used to make refractory bricks for furnaces. Soda ash, recovered from solid wastes, is used in the manufacture of cement. Calcium carbonate is used in the manufacture of tooth paste.

Low Energy Extrusion Technologies, a Canadian company, has come up with a plan of using industrial and domestic wastes in making products for daily use. The company uses fly-ash, clay, wire scrap and wood chips to make paving stones, tiles, chair legs and table tops. The company is running a thriving business using industrial and consumer wastes to manufacture advanced composite materials. Japanese scientists have turned raw sewage into a protein-rich product that resembles beef in texture. In this process, raw sewage is turned into protein-rich mud, to which soyabean, food additives and steak sauce are added to give the product an appetizing smell. The scientists believe that one day this "sludge-burger" might be used to feed livestock. According to researchers in Cotton College, Guwahati, the silk industry wastes containing large quantities of waste pupae can be used as poultry feed, because waste pupae of muga (a wild silk obtained from the cocoon of an Assamese moth) in dry form is very rich in animal protein. A mixture prepared from raw muga pupae and its deoiled form compares well with the conventional fish meal as a rich source of animal protein. It has been found that this feed significantly increases the growth rate of chicken.

MODEL QUESTIONS

(Essay/Long Type)

9.1 Describe in brief the sanitary land-filling method for solid wastes disposal.

9.2 Describe a modern composting method.

9.3 Write a critical note on biogas.

9.4 Describe a method to manufacture particle boards from wastes.

(Objective/Short Type)

9.5 What are social wastes?

9.6 What do you mean by "shredding the solid wastes"?

9.7 What is a green fuel?

9.8 Define biomethanation?

9.9 What is fly-ash?

9.10 What is incineration?

9.11 The highest calorific value can be obtained from:
 (a) Garbage
 (b) Rubbish
 (c) Pathological wastes
 (d) All have equal value.

9.12 The important vectors of human diseases in regard to solid wastes are:
 (a) Rats
 (b) Flies
 (c) Both
 (d) Only (b).

9.13 Aquatic weeds cause:
 (a) Water-born diseases
 (b) Evapotranspiration
 (c) Blockage of waterways
 (d) None of these.

9.14 Pyrolysis of plastics may yield:

    (a) Wax   (b) Grease

    (c) Adhesive   (d) All of these.

9.15 Match the following:

    (A) Anthracene   (a) Coal tar

    (B) Hydrometallurgy   (b) Electroplating

    (C) Polymerized wastes (c) Metal extraction

    (D) Silver   (d) Nonbiodegradable

9.16 Match the following:

    (A) Cement   (a) Biomass

    (B) Compost   (b) Fly-ash

    (C) Gasifier   (c) Dano process

    (D) Magnesium   (d) Refractory

9.17 Write true or false:

Open dumping of solid wastes is extensively practiced in India.

9.18 Write true or false:

Sanitary land-fill is an aerobic process.

9.19 Write true or false:

Composting is an anaerobic process.

9.20 Write true or false:

Generally, the agricultural wastes are non-hazardous.

9.21 Fill up the blank:

_____, obtained from corn cobs, is the basic chemical used for synthesis of nitrofurans.

9.22 Fill up the blank:

Land-fill operations are quite difficult to carry out during the _____season.

9.23 Fill up the blanks:

One can recover _____ and _____ from pickle liquor.

9.24 Fill up the blank:

Pathological wastes have a moisture content of about _____%.

9.25 Write the odd ones:

    (a) Nitrogen   (b) Oxygen

    (c) Phosphate   (d) Potash.

9.26 Write the odd ones:

    (a) Slurries   (b) Strong acids

    (c) Solids   (d) Sludges.

Sustainable Development

10.1 DILEMMA OF DEVELOPMENT

10.1.1 Ecological Aspects

The concern for the ecological imbalance and the consequent deteriorating environment, once assumed to be a luxury problem of the affluent nations, is now rapidly assuming a menacing aspect especially in developing countries. On the earth, man is a part of the natural phenomenon consisting of the soil, air, water, plants and animals (including man himself). Nature maintains a balance in the world ecosystem through the interactions of the above elements. Nature maintains its own purifying system in which plants and animals play a major role. Problems began when the world ecosystem was violently disturbed by short-sighted activities to benefit the community. The ecological equilibrium has been disturbed in many ways by man's endeavors to improve his living and to discover the secrets of nature.

10.1.2 The Paradox of Development

Unfortunately, development has been equated with economic growth and consumerism and the absence of industrialisation seems to carry a stigma of backwardness. Rapid industrialization has come to be believed as the only road to prosperity. Technological advances of the modern age have also brought about a process of "de-humanisation of man" and the diversity of human culture is being fast replaced by a sort of homogeneity in life. Indian civilisation, in particular, has always been a diversity of cultural heritage, even during the pre-historic days. In modern age, however, this diversity is under threat because technological progress offers no help in preserving the cultural diversity. Technology can give man materialistic comforts of life without caring for the preservation of life-support system or maintenance of balance in the world ecosystem. According to Morries Neiburger, an authority on environmental pollution, "All civilization will pass away not from a sudden cataclysm like a nuclear war, but from gradual suffocation in its own wastes". It has been estimated that more than 15,000 chemical compounds are widely used in industry, agriculture and commerce; but only a very small fraction of these has been thoroughly investigated to check their effects on human health.

10.1.3 Development and Environment

Environment and development are, in fact, two sides of the same coin. In the past, economic development and environmental issues were considered separately and the environment was treated by the economists as something external to the production process, which was difficult (and unnecessary) to quantify in terms of normal economic accounting. This attitude, however, is no longer justified and, in future, environmental impact of development projects has to be taken as seriously as the production targets.

The term *"environment"* is quite comprehensive. In a broader sense, environment is the sum total of all the external conditions and influences that affect the life and development of various organisms of the ecosystem.

One of the major issues which have reached the top of the political agenda pertains to energy. In recent years, industrialized countries have achieved considerable improvements in energy efficiency. However, the gains in this respect have been more than off-set by the corresponding growth in global energy consumption. It is now clear, therefore, that a much greater emphasis is required on energy conservation and energy efficiency in all sectors of human activity in developed as well as developing countries. Energy issues are closely linked with the problems of deforestation, transports, thermal and nuclear power plants and large dams. The transport infrastructure designed for the private vehicles has significant impact on the public transport sector. There has been reduction in investment in railways in almost all parts of the world, despite they being energy efficient and less polluting modes of mass transportation.

The Constitution of India contains specific provisions making it obligatory for the State to protect and improve the endowments of nature. However, inspite of the existence of several legislative measures for environmental protection, the implementation has been very tardy. The State Pollution Control Boards are not very effective due to lack of sufficient staff and resources and, as a result, there is a steady deterioration in the environmental quality. Indian industry, which is responsible for air and water pollution to a large extent, has so far given only lip service to pollution control. There is a general feeling that the industry is not fully willing to invest for protecting the environmental quality beyond the legal or statutory requirements. The environmental concerns which we face today have global implications. The community of nations has to act collectively for solving global problems.

10.1.4 Holistic Approach to Development

In order to cure the ills of the present concept of development, we should recognize the importance of holistic effects. In the language of ecology, the holistic view is known as the *"ecosystem view"*. For maintaining ecological balance in the world ecosystem, the whole system has to be managed, considering both the ecological and non-ecological segments. The holistic point of view may be clarified by a very simple example. A pebble when dropped into a pool of water, produces ripples in the form of concentric rings of waves forming around the point of impact and spreading outwards. It is only the water molecules at the point of impact of the pebble that are directly disturbed; but because of the forces of interaction that exist among water molecules, the disturbance is transmitted over a much larger area. The same holds true in case of ecologically sound development and other large-scale human activities. Environmental problems created by large development projects can be tackled successfully only when they are viewed as a whole.

In the Indian context, the dynamics of a village ecosystem offer a good example for the holistic approach. In a village, the farming system including both agriculture and animal husbandry, depends on natural vegetal resources i.e., grasslands and forests. Moreover, the demand for domestic energy is usually met from the forest; so forest plays a key role in the village ecosystem. Due to rapid increase in human and livestock population in India after the independence, the demand for energy has also increased and, as a result, the forests are being reduced at a fast rate. To take any kind of effective action to reverse an undesirable trend would require that we move as rapidly as possible to holistic approach to development. This requires, among other prerequisites, the development of holistic economics that integrates market as well as non-market values. A favorable future for mankind, thus, depends more on the integration of ecology and economics.

10.1.5 An Action Plan

Developmental activities cannot be totally stopped to save the environment. At the same time, it is possible to make the process of development safe by adopting suitable precautionary measures in line with nature's well-balanced ecosystems and by planning the environment side by side with the development planning. At the time of designing the development project, care should be taken to preserve all natural features such as hills, ridges, valleys, water fronts, ponds, springs, vegetation and wild life and to integrate them intelligently with the project design. The reason is that these natural features not only help in maintaining the delicate ecological balance and eliminating the pollution but also tend to reduce the mental tension so often experienced by people as a result of our fast life. As a result of innovative irrigation practices, improved dryland farming and controlled grazing, some deserts have now become productive and prosperous. In China, further deterioration has been halted in many desert areas and their productivity has been boosted up. The American Great Plains, once threatened to become a permanent dust bowl, are now productive and prosperous.

10.2 DEFINITION AND CONCEPT OF SUSTAINABLE DEVELOPMENT

10.2.1 Definition of Sustainable Development

In reality, all environmental problems are development problems. Many alternate development strategies have been propounded and they are being implemented without adequate environmental safeguards. In general, sustainability implies continuity of all things that are basically positive and might be thought of as broadly desirable or admirable. Some other definitions or interpretations of sustainability or sustainable development are as follows:

(1) Sustainability is the ability of an activity or development to continue in the long term without undermining that part of the environment which sustains it.
(2) Sustainable development is the development that seeks to improve the quality of human life without undermining the quality of our natural environment.
(3) Sustainability implies that human use or enjoyment of the world's natural and cultural resources should not, in overall terms, diminish or destroy them.

The term *Sustainable Development* was coined by the World Commission on Environment and Development (WCED), headed by Dr. (Mrs.) Gro Harlem Brundtland (the then Prime Minister of Norway). In its report *"Our Common Future"*, presented in 1987 (known as the **Brundtland Report**), sustainable development was defined as the "**development that meets the needs of the present without compromising the ability of the future generation to meet their own needs**". Sustainable development, therefore, implies protecting the environmental wealth, human capital stock, land, water and air, ecological living and non-living resources, and socio-economic resource base.

10.2.2 Concept of Sustainable Development (Historical)

The major landmarks in the history of sustainable development are as follows:

10.2.2.1 The Stockholm Conference

The United Nations Conference on Human Environment (commonly known as the "Stockholm Conference"), which was held in Stockholm in 1972, identified poverty as the main cause and the consequence of environmental degradation. The Stockholm Conference debated on earth ethic.

Barry Commoner, in his famous book *The Closing Circle* (1974), formulated four popular laws of ecology:

(1) Everything is connected to everything else.
(2) Everything must go somewhere.
(3) Nature knows best.
(4) There is no such thing as a free lunch.

Unfortunately, these laws were dubbed as "Principles of Impotence" and, therefore, they had only a limited impact on the affairs of the world. E.Cook, in his book *Man, Energy and Society* (1976), considered energy use, environmental sanity and quality of life as independent variables, each having at least one real limit.

10.2.2.2 The World Chapter for Nature

The World Chapter for Nature was adopted by the United Nations General Assembly in 1982. Later, the term "sustainable development" was well incorporated in the Report, "Our Common Future".

The Brundtland Report *"Our Common Future"* gives stress on the following:

(1) A political system for effective participation of citizens in decision making.
(2) An economic system to generate surpluses and technologies on a self reliant and sustained basis.
(3) A social system to resolve the social tensions.
(4) A production system to preserve the ecological base for development.
(5) A technical system to search new solutions continually.
(6) An administrative system which is flexible and capable of self-correction.
(7) An international system for faster sustainable patterns of trade and finance.

The Brundtland Report gives emphasis on the formulation of development strategies by all nations so as to achieve sustainable development. Sustainability would have to include the following:

(1) Sustainable environment.
(2) Sustainable world.
(3) Sustainable human development.
(4) Sustainable peace and development.
(5) Sustainable consumption.
(6) Sustainable technology.

The 1980's and 1990's witnessed several International Commissions and Conventions related to sustainable development. The prominent ones among these were the following:

(1) The Montreal Protocol on Ozone Depletion.
(2) The Basel Convention on Hazardous Wastes.
(3) The Climate Change convention.
(4) The Biodiversity Convention.

The Rio Earth Summit, held in Rio de Janeiro (Brazil) in June 1992, launched a global partnership on environment and development and adopted a declaration consisting of 27 principles with the objective of achieving sustainable development. Agenda 21, the Action Programme adopted at the Rio Earth Summit, devoted chapters relating poverty with environment and emphasized the importance of *Environmentally Sound Technologies* (ESTs), and the World Conference on Science laid further stress on these. The Copenhagen Declaration on Social Development (1995) stated that socio-economic development and environmental protection are interdependent components of sustainable development. The Cairo, Beijing and Istanbul conferences also came out with similar recommendations.

The *Human Development Report*, which was published in 1993, has termed the economic growth in recent years as jobless growth having adverse and dangerous consequences. In a report entitled *"State of the World"*, which was published in 1997, Lester Brown and his co-authors identified a group of Eight Environment Heavy Weights (commonly known as *"E8 Nations"*). The E8 Nations include four developed nations (USA, Russia, Japan and Germany) and four developing countries (China, India, Indonesia and Brazil). The E8 Nations jointly contribute 56% of the world's population, 58% of the world's carbon emissions, 59% of the world's gross product and 53% of the world's forest areas.

10.2.2.3 A Better Quality of Life

In 1999, Her Majesty's Government (HMG) of the United Kingdom published White Paper entitled *"A Better Quality of Life"*. This White

Paper identified 14 indicators for a better quality of life, which are as follows:

(1) The total output (Gross Domestic Product).
(2) The per capita investment in private, public and business assets.
(3) The percentage of people of working age in employment.
(4) Qualifications at age 19.
(5) The percentage of people with longer expected years of healthy life.
(6) The percentage of homes judged unfit to live in.
(7) The level of crime.
(8) The per capita emissions of green-house gases.
(9) The number of days in a year with air pollution moderate or high.
(10) The condition of road traffic.
(11) The quality and quantity of readily available water.
(12) The population of wild birds.
(13) The percentage of new homes built on previously developed land.
(14) The per capita waste generation and management of waste.

10.2.2.4 *World Summit on Sustainable Development*

The "World Summit on Sustainable Development", which was held at Johannesburg in August 2002, was one of the major technical events of the present century. At this Summit (also know as "Rio + 10 Summit"), representatives of more than 190 countries of the world discussed the sustainability problems being faced by mankind and tried to find the ways and means to mitigate the sufferings of common people living in developing countries. As part of the World Summit on Sustainable Development, the South African Government organized a "Forum on Science, Technology and Innovation for Sustainable Development" (known briefly as the "STI Forum"). Within the STI Forum, the World Federation of Engineering Organisations (WFEO) and the Engineering Council for South Africa (ECSA) organized a one-day session on "Engineering and Technology for Sustainable Development" on August 28, 2002. The main objective of this session was to highlight the crucial role of engineers in tackling the problems of sustainable development and poverty alleviation.

10.2.3 Sustainable Development and the Engineer

We are living now in a knowledge society where the innovations and applications of science and technology are vital for the eradication of poverty and for social and economic development on a sustainable basis. The engineers have a crucial role to play in poverty alleviation and sustainable development. Poverty, in simple terms, means the limited access of poor people to the relevant knowledge and material resources to meet their basic human needs like safe drinking water, sanitation, food, housing, transportation, communication, healthcare, income generation and recreation. The role of the engineering community in poverty alleviation includes advice, information, communication, project and programme activity (at the national, regional and international levels) relating to technology transfer, improved education and training, improvement of innovation systems, development of resources and human and institutional capacity building on engineering and technology.

Needless to say, the appropriate technology for poverty reduction has to be affordable and understandable by the poor people, preferably building upon the local knowledge, skills and materials, and delivering solutions that are technically viable, commercially feasible and socially and environmentally sustainable.

Through capacity building, education and improved engineering communication, engineers and technologists can empower a new generation of citizens with the technical knowledge they need to better understand and address the issues of sustainable development. Poverty is considered to be the single most important impediment to the sustainability of the quality of environment and life. Improving the economic well-being of developing countries and reducing their poverty is a prerequisite for creating a stable world and reaching the goal of sustainable development. Consequently, engineers should be of even greater help in the immediate future, than in the past, in achieving the goals of poverty alleviation and sustainable development.

Engineers can play a greater role in providing a better quality of life to our huge population provided they are committed to the "6 Rs":

(1) Reduce consumption.
(2) Reduce wastes.
(3) Replace unsustainable with sustainable.
(4) Reuse and recycle.
(5) Restructure institutions and infrastructures.
(6) Restore land, water and ecosystems.

In our mad rush into *Information Age* and *Knowledge Economy*, engineers must never forget the UNESCO Charter "we have not inherited the world from our forefathers but have borrowed it for our children". Unfortunately, most of the recent innovations in global science and technology have failed to help the poor and to preserve the environmental quality. For example, less than 10% of the current global expenditure on health research is directed towards the diseases of 90% of the people. That is why the renowned Indian scientist Dr. M.S. Swaminathan, while addressing the World Conference on Science (in June, 1999) said,

"Reaching the unreached and including the excluded have to be important components of the science and technology policy and strategy for the new century".

10.3 SCIENCE, TECHNOLOGY AND SUSTAINABLE DEVELOPMENT

10.3.1 Role of Science and Technology

For sustainable development, networking between policy makers, scientists, technologists, bureaucracy and development practitioners (i.e., the Planning Commision) is of paramount importance. Another important requirement for sustainable development is the availabity of environmental sound technologies. Environmentally sound technologies (ESTs) may be defined as follows:

"ESTs are the technologies that protect the environment, are less polluting, use all resources in a more sustainable manner, recycle more of their wastes and handle residual wastes in a more acceptable manner than the technologies for which they were substitutes".

The process of sustainable development has to be interactive, inter-sectoral, participatory and adaptive. It has, moreover, to be supported by other complementary measures like legal, financial, technical and institutional. Any missing link in this framework can undermine the process of sustainable development.

There is a wide-spread realization today that readymade solutions for sustainable development do not exist. Since sustainable development is a multi-disciplinary activity, we have to use all our ingenuity and combine the best of old practices with the latest innovations in science and technology to implement sustainable development projects. There are eight sectors which are most vital for sustainable development in India and other developing countries:

(1) Agriculture.
(2) Industrialisation.
(3) Infrastructure.
(4) Energy.
(5) Water.
(6) Healthcare.
(7) Pollution control.
(8) Biodiversity.

10.4 TECHNOLOGY FOR SUSTAINABLE ENERGY

10.4.1 Energy and Its Use

Energy may be defined as a property of matter which can be produced from or converted into mechanical work. Energy is a prerequisite for any development in today's world. The comforts and necessities of modern life cannot be possible without energy. Both energy production and energy utilization are, therefore, recognized as accurate indicators of a country's progress. We have a wide variety of energy sources such as fire-wood, animal and agricultural wastes, coal, oil, natural gas, electricity, nuclear energy etc. The sources of energy which are being continuously produced in nature are known as *renewable sources of energy*, e.g., wood, wind energy, tidal energy, solar energy, ocean current, nuclear fusion and biogas. On the other hand, the sources of energy which cannot be replaced if they are exhausted are called *non-renewable sources of energy*, e.g., coal, oil, natural gas, lignite and uranium. Thermal power, hydroelectric power and nuclear power are the main *conventional sources of energy*. The major *non-conventional energy sources* are solar energy, wind energy, geothermal energy, ocean current energy, tidal energy and biogas. In addition to environmental pollution, the world today is on the verge of yet another catastrophe, viz., the energy crisis. We need to develop a new jurisprudence that aims at sustainable development, which can act as a general guideline for the local, national and international policies for the management of energy resources.

10.4.2 Energy Conservation

Energy conservation is highly cost effective and, therefore, there is scope in sector of the economy to improve the efficiency of energy use. The Government of India has offered a variety of fiscal concessions to encourage energy saving measures by industry. Industrial Development Bank of India (IDBI) provides soft loans for such measures. With the deregulation and liberalization of the economy and more competition, energy conservation receives greater attention by the industrial sector. Moreover, there is plenty of rooms for energy conservation in other sectors, too, e.g., replacement of incandescent bulbs by fluorescent lighting in homes and more efficient suction and delivery systems in irrigation pumpsets. There is a large existing stock of obsolete equipment which consume relatively a larger amount of energy and the replacement of such out-dated equipment requires huge initial investment. Secondly, pricing of energy is a key factor in determining whether or not energy is efficiently used. In India, both coal and electricity have been under-priced. Until quite recently, the State Electricity Boards and the coal industry used to incur heavy losses. In some States, electricity is supplied almost free to farmers, while it is heavily subsidized in many other States. Indiscriminate subsidies have encouraged wasteful use of energy. Thirdly, the new technologies to harness non-conventional sources of energy (like wind power or tidal energy) are not yet capable of supplying energy on a large scale as needed for industry and other sectors of the economy.

Most of the renewable energy technologies are suited mainly for small-scale decentralized energy generation. For India, the most important forms of non-conventional energy are the solar energy, wind energy and biogas energy. In the past, energy conservation decisions have been adhoc and often inconsistent.

10.4.3 Strategies for Energy Sustainability

10.4.3.1 Short-Term Strategies

(1) To initiate measures for reducing losses during production, transportation and end-use of all forms of energy.

(2) To initiate research and development activities to reduce energy consumption in different sectors of the economy.

(3) To promote energy conservation by fiscal incentives.

(4) To discourage wasteful use of energy by proper pricing of energy.

(5) To initiate measures for meeting the basic energy requirements of the urban and rural households.

(6) To maximize the demand satisfaction for energy from indigenous sources (in order to reduce import of petroleum products).

(7) To maximize returns from the assets already existing in the energy sector.

10.4.3.2 Medium-Term Strategies

(1) To initiate steps towards progressive substitution of petroleum products by coal, lignite, wood and natural gas so as to restrict the quantum of oil imports.

(2) To accelerate the development of all renewable energy sources like solar energy, wind energy, biogas and the available hydro-electric potential.

(3) To accelerate research and development efforts on decentralized energy technologies based on renewable energy sources.

(4) To initiate appropriate organizational changes in different energy sub-sectors with a view to achieve energy sustainability.

10.4.3.3 Long-Term Strategies

(1) To enhance energy supply, based mainly on renewable sources of energy.

(2) To accelerate research and development activities with a view to increase the use of solar energy by making it more cost-effective.

(3) To promote technologies of production, transportation and end-use of energy that are environment friendly and highly cost-effective.

There are two major approaches to *Technology for Sustainable Energy*. In the first place, we can devise ways and means to use the currently available energy sources more efficiently and to reduce the waste of energy as much as possible. Secondly, we can develop renewable sources of energy and progressively replace the use of fossil fuels by using renewable energy in their place. In the following section, we discuss energy converters, which use the first approach and help conserve energy.

10.4.4 Energy Converters

10.4.4.1 Need for Energy Conversion

Often, we have to convert energy from one form to another for various reasons such as the convenience of energy transportation and use. It is very important to find the most abundant and environment friendly fuel and then to convert it into the most convenient form for transportation and use. This is done by using various types of energy converters.

10.4.4.2 Dual-Fuel Engines

From the point of view of environmental protection, natural gas is definitely superior to coal and oil. Accordingly, internal combustion engines and power stations are being designed to work on either oil or natural gas. The power utility industry already uses natural gas whenever possible. In New Delhi, buses have already switched over to compressed natural gas (CNG) as a fuel and this has led to considerable improvement in the quality of ambient air. The engines in other motor vehicles can also be modified easily to run on CNG. In case of CNG, the fuel economy is good, the wear on the engine is exceptionally low and the environmental pollution is also low. These advantages of CNG lead to a speculation that all internal combustion engines may eventually be compelled to use only natural gas as a fuel within city limits.

10.4.4.3 Hydrogen as Fuel

Oil is needed for the operation of our extensive transportation system. The transportation industry uses up to about 20% of our total energy supply, mostly in the form of petroleum products. There is no easy substitute for aviation fuels (light distillates of oil) for airplanes.

The only method now in use for the commercial production of oil is the combining of hydrogen and carbon monoxide mixture to produce hydrocarbons by catalytic synthesis. This method is used in South Africa. The possibility of the vanishing liquid fuels in the next few decades has stimulated the study of other fuels for the transportation industry. For example, we can produce hydrogen by the electrolysis of water and then use it in motor vehicles as a substitute for petrol or diesel. Unfortunately, hydrogen explodes easily; but if we can solve the problem of safety, hydrogen burns with very little environmental pollution since the residue is water vapour. However, nitrous oxide is produced if we use air (instead of pure oxygen) to burn hydrogen. An alternative to hydrogen gas as a fuel is lithium hydride, a solid, which may be produced electrolytically and which can be burned to produce energy.

10.4.4.4 *Methanol as a Fuel*

Methanol can be used instead of gasoline and fuel oils, or mixed with them, for heating and generating electricity. Methanol can be made from natural gas, coal, petroleum, vegetation, animal (and human) wastes and garbage. Excess production of associated natural gas from oil fields is hampered by the cost and difficulty of transporting the gas over long distances. Although the production of methanol from natural gas is less efficient and more expensive, methanol can be transported more easily and, at the final destination, it can be used directly or gasified back to methane gas (CH_4), depending on the market conditions. Methanol may be added to petrol up to about 10% by volume without necessary adjustment to an automobile engine. Of course, methanol and petrol separate at low temperatures, but nonpollutant-type additives can be used to prevent their separation. Pure methanol can be used in automobile engines but that requires a modified (or adjustable) carburetor. On the other hand, the addition of methanol to petrol or complete substitution of it by pure methanol, eliminates the need for lead additives as antiknocking agents. This results in cleaner air.

The chemical process for the conversion of coal or coke to methanol proceeds as follows:

$$C + 2H_2O \longrightarrow CO_2 + 2H_2$$
$$C + H_2O \longrightarrow CO + H_2$$
$$CO + 2H_2 \longrightarrow CH_3OH.$$

10.4.4.5 *Fuel Cells*

Fuel Cells are divided into two classes: (1) Primary cells and (2) Secondary cells. The ordinary dry cell, used in torch lights, transistor radios and calculators, is an example of a primary fuel cell. Its reactants are converted, during the chemical reactions, into electricity. The original chemicals are consumed. Secondary fuel cells, also called "storage batteries" or "accumulators", are best known in the form of lead-acid batteries as used in automobiles. They are reversible devices because they are charged with electricity and then discharge electricity. In this case, the original chemical materials are not consumed in the complete cycle of charging and discharging. A fuel cell that is practical for large-scale power generation was not realized for a long time due to the lack of understanding of electron transfer reactions across interfaces. However, modern fuel cells rapidly reached such a stage of perfection, efficiency and low weight that hydrogen-oxygen fuel cells have provided auxiliary power in American spacecrafts since 1966.

There are many types of fuel cells both in use and under development. The fuel used may be solid, liquid or gaseous. Metals, gases, hydrocarbons and even biochemicals are being explored as fuels for fuel cells. The basic fuel cell uses the reaction between hydrogen and oxygen to form water. The rapid combustion of hydrogen that we are most familiar with is an explosion giving off heat and light. In the fuel cell, however, the reaction is carried out in a controlled manner. In contrast to batteries, which are closed systems in the chemical sense, the fuel cells are open systems in the sense that they are constantly fed with fresh reactants and the products of reaction are constantly removed.

Hydrogen can be produced by electrolysis of water and then used to operate fuel cells in cities. The fuel cells would make a pollution free engine, although hydrogen gas is explosive and thus poses safety problems that would have to be solved. More interesting would be the use of propane.

10.4.4.6 *Photovoltaic Cells*

Photovoltaic energy conversion is a direct energy conversion in the sense that light energy is converted directly to electrical energy without intermediate involvement of mechanical or thermal energy (i.e., the use of steam turbine or generator). The efficiency of photovoltaic converter is limited by the Carnot cycle to the extent that the source has a temperature equal to that of the surface of the sun (about 6,000° K), which leads to high efficiency. Photovoltaic energy converters are more commonly known as *"solar cells"*. A photovoltaic (or photoelectric) energy converter consists of a semiconductor p-n junction. When the solar cell is exposed to sunlight, a photon (a small bundle of light energy) transfers its energy to an electron when the two collide. The electron may then have sufficient energy to become free. Such electrons, released by solar photons by the process of photoelectric effect, can be made to flow. The maximum predicted efficiency for solar cells is of the order of 20%. Since sunlight is free, the fuel cost for a solar cell is zero. Solar cells have been extensively used for space applications. It is necessary, however, to couple solar cells with storage batteries to ensure a continuous supply of electricity.

10.4.4.7 Magnetohydrodynamic (MHD) Energy Converters

As long as we employ heat engines, the search will continue for the high-temperature working fluids necessary to obtain a high efficiency in the Carnot cycle. One such possibility is a magnetohydrodynamic (MHD) plasma, which is basically a mixture of positive and negative ions that move through a magnetic field in a direction perpendicular to the field. Under these conditions, an electric field is induced in a direction perpendicular to both the magnetic field and the direction of motion of the particles. As a result, the positive ions and negative electrons are directed towards opposite electrodes, giving rise to an electric current through a suitable load. Thus, in a magnetohydrodynamic energy converter, the translational kinetic energy of the positive and negative ions is converted into electrical energy. The efficiency can be as high as 50-60%. The gas that leaves the energy conversion chamber is at a temperature high enough to be used to generate steam. The steam produced in this way may then drive a turbine attached to a generator to produce electricity.

10.4.5 Renewable Energy Sources

10.4.5.1 Solar Energy

Solar energy is the primary source of all energy forms on earth with the possible exception of nuclear energy. The present-day interest in solar energy has arisen mainly due to rapidly dwindling of other energy sources. Solar energy became popular towards the end of the nineteenth century, when a number of solar-powered pumping stations and desalination plants were constructed.

The sun produces an enormous amount of energy through a series of thermonuclear reactions. Due to absorption and scattering of sunlight in the atmosphere, some energy is lost and the maximum flux of solar energy on the surface of the earth is about one kilowatt per square meter ($1kW/m^2$). The amount of solar radiation received at a particular place depends upon the time of the day, season, weather and the latitude of the place. India is fortunate to receive abundant solar energy with about 250-300 days of useful sunshine in a year. This enormous amount of solar energy may be converted into useful form of energy either through photovoltaic or thermal devices.

A major drawback of solar energy is that it is in a highly diffused form, unlike coal and oil which contain concentrated energy. When coal, oil or natural gas is burned, it can produce heat energy at high temperatures, which can then be converted into mechanical or electrical energy with a high degree of efficiency. Since the solar energy is diffused, it is not possible to convert the heat contained in sunlight into other forms of energy with great efficiency. As a result, the most efficient method of using the solar energy is to use it directly in the form it is received, i.e., heat. Solar thermal applications may be classified into two categories. The first category consists of applications at relatively lower and medium temperatures for heating water and air, cooking, drying and desalination. The second category includes application at higher temperature to produce steam for deriving mechanical or electrical power. A flat-plate collector system is suitable for lower temperature applications of solar energy, while a system concentrating the sunlight is suitable for high temperature applications. Solar cookers, water heaters, stills, kilns and dryers are based on flat-plate collector systems. Depending upon the design, solar concentrating systems are capable of producing temperatures upto 2000°C. These systems are best suited to produce low-pressure system for a number of industrial applications. A parabolic dish reflecting sunlight can concentrate the heat energy of sunlight at its focus, producing temperature upto 2000°C.

Solar cell is the most elegant method of converting solar energy into electricity. Solar batteries (systems consisting of a large number of solar cells) can operate all electrical devices for lighting, domestic applications, telecommunication, signaling, water pumping, etc. Some industries (like metallurgical and cement industries) require heat at very high temperature; but there are a large number of other industries (like paper, textile, pulp and pharmaceutical industries), which require heat at temperature below 500°C. Even if a part of the heat required by industries can be produced from solar energy, a considerable amount of fossil fuels can be saved and the quality of ambient air can be enhanced.

10.4.5.2 Hydroelectric Energy

Hydroelectricity is generated by converting the potential energy of water stored in a dam constructed across a river into electricity with the help of water turbines attached to generators. Hydroelectricity is playing a very important role in saving the environment since it does not contribute to air pollution or emission of green-house gases as the thermal power plants do. At present, hydroelectricity accounts for about 20% of the electricity generated in the world. India has a large hydroelectric potential of about 90,000MW at 60% load factor. Much of this huge estimated potential is yet to be tapped. The reason is that, with the rapid increase in the demand for power after the independence, priority has been given in India for the installation of super thermal power stations because their completion time is much shorter than that for hydroelectric power plants. However, hydropower stations are more economical for power generation and, with modern technology, their gestation period also can be reduced considerably to 2-4 years.

Some of the major advantages of hydropower stations over other types of power plants are:

(1) Hydropower plants have a longer life.

(2) Hydropower plants require minimum operating staff and less maintenance cost.

(3) As per the requirements of grid conditions, they can be very quickly started and stopped; so they are more suitable for varying loads.

(4) The efficiency of hydropower plants is much higher than that of thermal power plants.

(5) These units can be operated in a wider range of frequencies.

(6) These plants save scarce fossil fuels.

(7) They are non-polluting and environment friendly.

(8) Hydropower is a renewable and reliable source of energy.

(9) The cost of power generation including operation and maintenance using a hydropower plant is much lower compared to that for other sources of energy.

(10) Cost of power generation using hydropower plants is free from the inflationary effects after the initial installation.

(11) Reservoir-based hydropower projects often provide the benefits of irrigation, flood control, drinking water supply, navigation and recreation.

(12) Being labour-intensive, hydropower projects provide employment opportunities for the local people during the construction period.

In brief, hydropower is the cleanest, cheapest and the best (from the viewpoint of resource optimization) among all the sources of power generation.

10.4.5.3 Wind Energy

Wind energy is fast emerging as the most cost-effective source of power because it combines the abundance of a natural source (the wind) with modern technology. Since the energy crisis of 1973-74, the interest in wind energy has increased throughout the world. The most important advantage of wind energy is that it is produced without causing any harm to the environment. A wind-driven power station consumes no fuel or other raw materials and does not emit any waste gases or other waste materials. Wind energy has already become the world's fastest growing renewable source of energy. Wind power is of great significance in India since there are large coastal, hill and desert areas where wind energy can be gainfully exploited for the generation of electricity and pumping of water.

The present-day technology for harnesing wind power can be broadly divided into the following three categories:

(1) Wind pumps, which use the mechanical energy derived from wind power.
(2) Wind electricity generators, which convert the wind energy into electrical energy with the help of turbo-generators.
(3) Wind battery chargers, which produce electricity and store it in batteries.

In India, the exercise to harness wind energy includes wind pumps, wind electricity generators and wind battery chargers. In our country, there are many areas which are quite windy. Average annual wind intensity of about $3kW/m^2/day$ is prevalent at a number of places in Peninsular India and also along the coastline in Gujarat, Western Ghats and some parts of Central India. The wind intensities are even more than $10kW/m^2/day$ during winter and wind intensities exceeding $4kW/m^2/day$ are available for 5-7 months in a year.

The technology for harnessing wind energy is quite simple. When the blowing wind strikes the specially designed blades of a rotor, it causes the rotor to rotate. This rotation contains mechanical energy, which is derived from the kinetic energy of the wind. When the rotor is coupled to a generator, the mechanical energy of the rotor is converted into electrical energy. The Indian subcontinent is a high-wind zone with wind energy potential estimated at about 20,000 MW. So far, wind energy has been utilized mostly in Tamil Nadu, Andhra Pradesh and Gujarat for pumping water in rural areas. Germany, Denmark, the US, India and Spain have emerged as world leaders in the development of wind energy. With technological advances, the cost of generating wind power has continued to decline and wind promises to become a major source of power during the current century. The Washington-based World Watch Institute (WWI) has recognized India as a Wind Superpower in the world. In terms of total installed capacity, India ranks fourth in the world after Germany, Denmark and the US.

10.4.5.4 Geo-Thermal Energy

Geo-thermal energy is the heat energy of rocks lying deep within the earth. The temperature of earth's interior increases with depth. At most sites, however, the temperatures suitable for the utilization of geo-thermal energy on economic basis are found only at considerable depths. A temperature of 300°C normally occurs only at a depth of about 10km. In volcanic zones, temperatures of 300°C often occur at a lesser depth. The heat in deep rocks then reaches the surface through natural circulation of ground water. Modern technologies have made it possible to tap this source of energy for various commercial purposes. Worldwide utilization of geo-thermal energy is growing at a rapid pace, but India is far behind.

In India, there are more than 340 thermal springs with strong potential for the development of geo-thermal energy. Many of these are located in the West Coast, the Narmada rift and the Himalayan region. The surface temperature of these thermal springs ranges from 37°C to 90°C. In addition to generating power, the heat from these springs can also be utilized directly for other purposes. For example, the thermal springs along the Himalayan belt can be utilized for space heating. Other important geo-thermal areas in the country are Cambay Basin in Gujarat, Alaknanda Valley in Uttaranchal and Parvati Valley in Himachal Pradesh. A number of geo-thermal wells have been drilled in Ladakh to harness the geo-thermal energy in this area. The scientists from the Regional Research Laboratory, Jammu have used the geo-thermal energy in Ladakh for extraction and refining of salt, sulphur and borax in the Valley.

Efforts are being made to use the geo-thermal energy for generating power and for the purpose of refrigeration. Geo-thermal energy can either be used as steam to generate electricity in power stations or it can be used directly as primary heat. Geo-thermal energy is fast emerging as a significant source of electricity in several island nations of the world, mainly in the Indian Ocean and the Pacific region. Indonesia is setting up two geo-thermal power plants (each of 55 MW capacity) in the island of Java. In New Zeland, geo-thermal energy accounts for 8% of the country's installed power capacity. In Iceland, geo-thermal energy provides heating and hot water for more than 85% of the houses. France and Germany are also using geo-thermal energy for domestic heating.

10.4.5.5 Ocean Energy

The various sources for extracting energy from the ocean are as follows:

(1) Ocean Winds.

(2) Ocean waves.

(3) Ocean tides.

(4) Ocean currents.

(5) Ocean geo-thermal energy.

(6) Ocean thermal energy conversion.

(7) Ocean salinity gradient.

(8) Bioconversion of sea weeds.

These are very briefly described below:

(1) Ocean Winds: Coastal areas generally have stronger winds and, there-fore, enough energy is available from the wind. Meteorological data show that the average speed of wind in the lower troposphere is about 10 m/s, and

the available wind power is of the order of 500 W/m^2. Suitable designs are available for wind mills with battery bank systems for storing electrical energy.

(2) Ocean Waves: Ocean waves keep the water under constant motion. This continual motion can be harnessed to produce energy. The vertical rise and fall of successive waves is used to activate either an air-operated turbine or water turbine. The stronger the wave action and the wave height, the more is the power generated. The Ocean Engineering Centre at the Indian Institute of Technology, Chennai has developed an indigenous design for power generation using ocean waves.

(3) Ocean Tides: The use of ocean tides is the most feasible method for utilizing the ocean energy. Gravitational forces of the sun and the moon cause regular flow and ebb-tides in oceans. Power can be generated if the difference between the high and low tides is significant and water storage facility is available. The incoming tide is allowed to flow into the reservoir through a dam provided with turbines and the outgoing tide is also allowed to flow through the same dam to turn the water turbines for generating electricity. Tidal energy is important for India because it is renewable, pollution free and more stable compared to hydroelectric power.

(4) Ocean Currents: A number of designs are available to convert the energy of the fast-moving ocean currents into electrical energy. Water is allowed to pass through a series of turbines installed under water. However, it is more difficult to harness the energy of ocean currents compared to that of ocean tides because the energy density (the amount of available energy per unit volume of water) is fairly low in ocean currents.

(5) Ocean Geo-Thermal Energy: Certain areas of the ocean contain hot springs with temperatures as high as 50° C. It is possible to extract energy from such a temperature difference. Heat exchangers have been developed to establish a cycle for generating power. However, this renewable source of energy has only a very limited value at present.

(6) Ocean Thermal Energy Conversion: The principle of ocean thermal energy conversion is very simple. A working fluid having a low boiling point (such as ammonia or propane) is pumped into a closed tube exposed to warm water, which vaporizes the working fluid. This vapour condenses to give back the fluid in a liquid form at the cold water zone. If the vapour is allowed to pass through a turbine, it can turn the turbine to generate power. The working fluid is pumped up and made to circulate to get energy.

(7) Ocean Salinity Gradient: The principle for tapping energy from ocean salinity differences is based on osmosis. The movement of ions can cause an electric current to flow in water. This method uses the concentration

gradients of salts in oceans to generate power. In Sweden, work is in progress to generate 200 MW of power by utilising ocean salinity gradients.

(8) Bioconversion of Sea Weeds: In addition to various other methods, a very simple method is available to convert sea weeds into fuel and that is bioconversion. Sea weeds trap solar energy by photosynthesis; so they can be made to yield food and fuel. Easily harvestable sea weeds can be cultivated on nylon and coconut ropes and lines. These lines are then transported to the processing plants, where the sea weeds are clipped without harming the mother plants. The lines are then returned to the sea. In the processing plant, the sea weeds are converted into methane, food and fertilizers with the help of micro-organisms.

10.5 TECHNOLOGY FOR SUSTAINABLE MATERIALS

10.5.1 Waste Utilisation

Nothing, in fact, is really a waste. Today's waste may be regarded as tomorrow's raw material. Now it is almost certain that a nation that will not be able to recycle used materials and wastes will not be able to sustain itself. There is a global realization of the fact that single use of some of the important materials like metals, glass and paper would lead to scarcity of such materials since their feed-stocks would get exhausted. In developed countries, resource recovery is a high technology area. In many developing countries like India, however city garbage is sorted out manually for the recovery of metals, glass, plastics, paper, etc., which generate employment and materials for reuse.

The reasons for waste utilization are: (1) economic (2) environmental (3) resource conservation (4) employment generation and (5) provision of basic necessities of life. Developing countries with constraints of limited resources can not afford to call any material as waste. Paradoxically, however, it is the developed countries like Japan, Germany, the US and the UK that are in the forefront of waste utilization.

Wastes, if allowed to remain unutilized, can be potential environmental hazards in spreading diseases and leaching harmful chemicals into life-support systems. Resource conservation is yet another advantage that results from waste utilization. Furthermore, the decentralised management of wastes can generate employment by way of collection, handling, transportation and processing of wastes. In addition, many useful products can be derived from waste utilization, which are of value as food, fuel, fertilizer, fibre, paper, pulp, industrial chemicals, plastics, glass and metals. An obvious constraint on waste utilization is the dispersed nature of wastes, which makes their collection expensive. This problem can be solved by public participation and cooperative movement. In order to encourage industries to take to waste utilization, there should be fiscal incentives for using recycled materials and penalties for not recycling their wastes.

A crash programme for technology generation in the case of high-priority items need to be undertaken. In fact, there is a need for a central coordinating agency to give momentum to waste utilization programmes. This agency should take care of all aspects of the problem right from technology generation, engineering designs to complete recycling plants on a turn-key basis. One of the major constraints in waste utilization is lack of research and development work on waste recycling, which is not a fashionable area of research. There is also a general lack of technological availability and reliable feasibility reports. Further, accurate information on the availability of industrial, agricultural and municipal wastes, both location-wise and season-wise, are not available. The other major problem is the lack of a nodal agency to look after this inter-disciplinary area.

Technology for sustainable materials is described below:

10.5.2 Recovery of Metals

10.5.2.1 Need for Metals Recovery

Metals have played an essential role in the development of our society. In the fulfillment of this role, recycling has become almost indispensable. The importance of metals recovery arises from the following factors:

(1) Considerable market value of non-ferrous metals.
(2) Ease of recycling to give a material often indistinguishable from the primary material.
(3) Indestructibility of metals.
(4) Lower cost of recycling compared with primary production.
(5) Lack of indigenous resources.
(6) Political instability of primary metal producing countries.
(7) Economic pressure from within the industry.
(8) Technical requirements.

Appreciation for the need to recycle metals received considerable impetus from the publication of *The Limits to Growth* by D.K. Meadows and co-authors in 1972. These authors predicted growth rate and projected lifetime of a range of mineral resources. According to them, almost no primary source of metals will be left after about 100 years if the present trend of growth continues. The value of metals as scrap is roughly half the value of virgin material; but chemical wastes containing metals have a significantly lower value because of the need for costly concentration and purification. As in primary metal production, the cost of extraction increases as the metal content decreases. Future development of secondary metal industries (i.e., industries recovering metals from industrial and municipal wastes) depends on the following:

(1) More efficient and thorough reclamation of scrap.
(2) More efficient recycling processes.
(3) Product design with a view to recycling.
(4) Economic optimization of alternate recycling technologies.
(5) Changes in industrial and consumer attitudes.

Progress has already been made with the development of a wide range of physical separation methods for municipal wastes and shredded vehicle scraps. With the growth of hydrometallurgy for primary metal extraction, a number of operations and complete processes have been developed for the chemical processing of scraps and metal-bearing materials. Recovery of some important metals is briefly described below:

10.5.2.2 *Aluminium*

Of all the non-ferrous metals, aluminium offers the greatest inducement for recycling since the growth rate of aluminium is much higher than that for other metals. On a world-wide basis, about 27% of aluminium is produced from recycled scrap. The two basic sources of secondary material are classified as: (1) essentially new scrap arising from manufacturing processes; and (2) old scrap, arising from obsolete products.

Aluminium scrap requires sorting and some pre-treatment to prepare it in a form suitable for charging to the smelter. There are two types of refining operations: (1) removal of non-metallic contamination mainly aluminium oxide and (2) removal of undesired alloying elements like magnesium, sodium and calcium. Aluminium oxide forms immediately when the metal contacts air; so smelting of scrap metal must be carried out with the minimum contact with air. Molten salt fluxes are normally used and oxide collects in the flux. Hydrogen is also removed as a consequence of fluxing. Magnesium is removed as oxide, chloride or fluoride.

10.5.2.3 *Copper*

Copper is one of the oldest known metals and it has been recycled since the Bronze Age. Copper combines valuable mechanical and electrical properties with unique aesthetic appeal. Copper scrap contributes about 40% of the total production of copper and this percentage is increasing as virgin supplies are diminishing. There is a wide variety of copper-bearing wastes and scrap ranging from high-grade electrolytic scrap that only requires remelting or electrolysis, down to low-grade drosses and chemicals which require full smelting and refining operations. About half of all copper produced go to electrical industry, and insulated cable is the largest single source for secondary copper. Brass, bronze and gun metal are significant sources of secondary copper, as are copper products from the transport, building and other industries. The secondary copper industry relies mainly on pyrometallurgical processes to smelt and refine the metal and alloys. Some materials require pre-treatment after sorting, the most important example being electrical cable. The two main processes to separate copper from the covering material are: (1) burning and (2) granulation or shredding. Burning is relatively cheap, but it results in air pollution (smoke and hydrogen chloride from PVC). Shredding is followed by mechanical separation of the metal. The fundamental operation in secondary copper refining is the reduction of copper oxides to copper, together with the removal of impurities like zinc, aluminium and magnesium as a slag. Reduction is done in a blast furnace by adding coke. The "black" copper thus produced is impure (about 74% copper). It is refined in a converter for oxidation of impurities.

10.5.2.4 Iron and Steel

Almost all scrap iron and steel are processed pyrometallurgically, although serious attempts have also been made to employ hydrometallurgy. Modern basic steel-making technology requires a certain level of scrap for cooling during oxygen blowing (at a level of about 30% of the molten iron charge). The other significant scrap processing operation is the electric arc furnace, which uses 100% scrap feed. The principle of chemical recovery of metals is attractive for the following reasons:

(1) Low-grade and low-cost scrap may be employed.
(2) Good separation may easily be obtained.
(3) High purity of products is usually obtained by chemical methods.
(4) Energy consumption can be significantly less as compared with conventional methods.

10.5.2.5 Other Metals

All metals are recycled to a certain extent. In general, the more valuable a metal, the greater the degree of recycling. Each metal (and sometimes each industry) has its own technology for recycling, which can be very specific. The techniques are often confidential and economics dictate what and how much has to be reclaimed. Almost 100% recycling is obligatory in certain specialist applications involving radioactive substances. Good recovery is also desirable of those metals which are particularly toxic (like mercury, antimony, arsenic, cadmium and beryllium).

10.5.3 Recovery of Non-Metals

10.5.3.1 Important Non-Metallic Materials

Industrial, municipal and agricultural waste contain many important non-metallic materials:

  (1) Paper.
  (2) Glass.
  (3) Plastics.
  (4) Crop residues.
  (5) Animal dung.
  (6) Rubber.
  (7) Textiles.
  (8) Wood.
  (9) Mineral wastes.
(10) Other wastes.

Recycling and use of paper (Re: 9.3.3.2.), glass (Re: 9.3.3.3.), plastics (Re: 9.3.3.5.), crop residues (Re: 9.2.3.1.) and animal dung (Re: 9.2.2.3.) have been described in earlier chapters. Recycling and use of others in brief are described below:

10.5.3.2 Rubber

The single largest application of rubber is for vehicle tyres, which consume about 50% of the total rubber produced in the world. The other important uses of rubber are in cables, hose and tubing, belting, footwear and cellular products. Except for tyres, no other uses of rubber present a disposal problem, nor have a significant potential for recovery (except for new scrap and in-plant recycling). Two products can be obtained from scrap tyres by physical treatment: (1) "crumb" and (2) "reclaim". Crumb is a granular material produced from the peripheral surface of the tyre during buffing to prepare a clean straight surface to receive the new tread. Crumb can be reincorporated into new rubber compound to the extent of about 10%. Reclaim is essentially devulcanised rubber, and is processed and sold in sheet form. The reclaim rubber sheet is suitable for blending with new rubber to the extent of 30-40%. The separated steel wire is sold as low-grade scrap metal.

10.5.3.3 Textiles

Recycling, reprocessing and reuse of fibre wastes are traditional in the textile industry. When natural fibres are present in wastes, they can be reprocessed by a relatively simple operation and then blended directly with virgin fibres for reuse. Reprocessing has become more complicated now by the increased use of natural man-made fibre blends. This has led to a decline in the textiles reclamation industry. Textile wastes are collected by a network of merchants and passed on to reprocessors. Processing reduces the waste to fibres, which are blended with virgin fibres, carded and spun into thread and yarn. Essentially, the wastes are sorted by hand according to the shade and type. They are then "opened", i.e., reduced to fibres by a rag-pulling machine. Man-made fibres that are not contaminated can be reprocessed for higher-grade applications. Noncontaminated thermoplastic man-made fibres (nylon, polyester and polyolefins) can be recovered by melt processing.

10.5.3.4 Wood

In addition to their use in pulp and paper manufacture, chipped wood and sawdust may be used to supplement fuel in the factory where they arise. Sawdust is an adsorbent and is used as a filler in synthetic resins. Chips are used in chipboard manufacture, a product so successful that chips are now produced specially for it. Further, it is practicable for sawdust or wood chips to be converted to fermentable sugar by hydrolysis; but this is not yet economical.

10.5.3.5 Mineral Wastes

Some of the important mineral wastes are: (1) colliery spoil (2) china clay waste (3) slate waste (4) flyash (5) red mud (6) calcium sulphate and (7) blast furnace slag. Although these are generally of low intrinsic value, many of them have the potential for reuse in the building and construction industries.

Approximately half of the material extracted in coal mining is waste or spoil and most of it is disposed of by dumping in heaps. Minerals present in coal spoil comprise quartz, clay minerals, pyrites, and carbonates of calcium, magnesium and iron. Colliery spoil is used as fill in road embankments and building sites. Light weight aggregate is made from colliery spoil for use in the manufacture of precast concrete blocks.

China clay is produced from granite rock for use in the ceramic industry. About 80-90% of the extracted rock is spoil, which includes over burden, waste rock and coarse sand. The bulk of these materials is dumped. Coarse sand is used in road construction and in the manufacture of building materials.

About 90% of the output from slate quarries results in slate waste. Ground slate is used as an inert filler in plastic and rubber products and as granules in roofing felt. Light weight aggregates can be produced from slate waste.

Huge amounts of flyash are produced by coal-fired thermal power plants, steel plants and other industries. Some of the important uses of flyash are:

(1) fill for roads and building sites (2) concrete blocks (3) precast concrete (4) lightweight aggregate (5) cement manufacture (6) fired clay bricks (7) industrial filler (8) cement-stabilised soil and (9) concrete mortar.

The alkaline extraction of alumina produces large quantity of residues, which are known as "red mud" because of their brick-red colour due to the presence of iron oxide. Around one-third of this waste consists of ferric oxide. Considerable research has been conducted with the aim of recovering iron and some of the other constituents of red mud, or using it in some other ways. The iron recovery processes are mostly based on a complex multi-stage smelting procedure; but they are not economically viable. The red mud has also been used in clay products, both as a colouring agent and as a replacement material. It is an efficient adsorbent for sulphur compounds in stack gases and effluent treatment. Red mud may also be used as fill, for soil stabilization, as a catalyst for coal hydrogenation and to produce a chloride-resistant cement.

Calcium sulphate is produced as a byproduct of industrial processes (mainly phosphate and phosphoric acid manufacture, hydrofluoric acid manufacture and neutralization of waste sulphuric acid). The major outlet for calcium sulphate is as plaster of paris in the building industry. Other uses for waste calcium sulphate are as a filler for polymeric materials, as a soil conditioner and as a source of sulphate for the manufacture of sulphuric acid or ammonium sulphate.

Blast furnace slag, a byproduct of iron manufacture, is a waste which is a well- established construction material. Selective cooling results in the production of three different types of slag:

(1) A crystalline product, resembling igneous rock, produced by slow cooling in air (suitable for use as roadstone or dense aggregate for concrete).

(2) A lightweight aggregate, produced by foaming molten slag with a limited amount of water (used for block making and for structural reinforced concrete).

(3) Quench glass, produced by rapid cooling (used in the manufacture of slag cements).

In addition, blast furnace slag is converted into a synthetic fibre for slag wool.

10.5.3.6 *Other Wastes*

There are many other wastes, both organic and inorganic, for which recycling is economically feasible. As a general rule, a material which is not over-degraded in the manufacturing process may either be directly recycled or reused in-house for a product of different specifications. For example, pottery waste can often be reground and reused for a lower-specification product and waste sanitary porcelain can be reused in ceramic tile manufacture.

For inert inorganic solid wastes, there may be an outlet as a filler, for example in resin bonded formulations or cements. A common problem, however, is the cost of transport from the producer to the consumer. Some metallurgical slags have special properties and are reused in limited amounts. Copper blast-furnace slag is a valuable grit-blasting material, while reject tin slags are used for road-making or grit-blasting.

Organic wastes invariably have the potential for thermal treatment to produce energy, fuel or organic chemicals. Food-processing wastes may form a valuable source of animal foodstuffs. They may be recycled directly (as pig food) or indirectly, following drying and reinforcement with protein and vitamins. Alternatively, food wastes can be hydrolysed and fermented to produce alcohol. Compost of organic wastes can be used as a fertilizer.

MODEL QUESTIONS

(Essay/Long Type)

10.1 Define and explain sustainable development.

10.2 What are the salient features of the Brundtland Report, "Our Common Future"?

10.3 What are the fourteen indicators for "A Better Quality of Life" as contained in 1999 White Paper of the Government of UK ?

10.4 Write briefly on the Technology for Sustainable Energy.

10.5 Write a critical note on the Technology for Sustainable materials.

(Objective/Short Type)

10.6 What is CNG and what are its main uses ?

10.7 Write all the steps of chemical reactions for conversion of coke or coal to methanol.

10.8 What is a photovoltaic cell ?

10.9 Give examples of four non-metallic materials which can be recycled and reused.

10.10 What are the uses of china clay ?

10.11 At the time of designing a development project, care should be taken to preserve:

    (a) Hills   (b) Valleys

    (c) Springs   (d) None of these.

10.12 Barry Commoner, in his famous book "The Closing Circle," formulated:

    (a) Two laws of ecology   (b) Four laws of ecology

    (c) Three laws of ecology   (d) No law of ecology.

10.13 Sectors which are most vital for sustainable development of India are:

    (a) Agriculture   (b) Infrastructure

    (c) Water   (d) All of these.

10.14 Match the following:

    (A) Energy   (a) Ozone depletion

    (B) Geo-thermal energy   (b) Our common future

    (C) Brundtland   (c) Transport sector

    (D) Montreal protocol   (d) Cambay Basin

10.15 Write true or false:

Nothing, in fact, is really a waste.

10.16 Write true or false:

Primary metal industries recover metals from municipal wastes.

10.17 Name the odd ones:

    (a) Glass   (b) Plastics

    (c) Aluminium   (d) Rubber.

10.18 Name the odd ones:

    (a) Hydrogen   (b) Methanol

    (c) Coal   (d) Iron.

10.19 Fill up the blank:

All natural features maintain _____ balance.

10.20 Fill up the blanks:

We have _____ inherited the world from our forefathers but have borrowed it for our _____.